AF802470

Models for Multispecies Management

Contributions to Economics

Christoph M. Schneider
Research and Development Management: From the Soviet Union to Russia
1994. ISBN 3-7908-0757-5

Bernhard Böhm/Lionello F. Punzo (Eds.)
Economic Performance
1994. ISBN 3-7908-0811-3

Lars Olof Persson/Ulf Wiberg
Microregional Fragmentation
1995. ISBN 3-7908-0855-5

Ernesto Felli/Furio C. Rosati/Giovanni Tria (Eds.)
The Service Sector: Productivity and Growth
1995. ISBN 3-7908-0875-X

Giuseppe Munda
Multicriteria Evaluation in Fuzzy Environment
1995. ISBN 3-7908-0892-X

Giovanni Galizzi/Luciano Venturini (Eds.)
Economics of Innovation: The Case of Food Industry
1996. ISBN 3-7908-0911-X

David T. Johnson
Poverty, Inequality and Social Welfare in Australia
1996. ISBN 3-7908-0942-X

Rongxing Guo
Border-Regional Economics
1996. ISBN 3-7908-0943-8

Oliver Fratzscher
The Political Economy of Trade Integration
1996. ISBN 3-7908-0945-4

Ulrich Landwehr
Industrial Mobility and Public Policy
1996. ISBN 3-7908-0949-7

Arnold Picot/Ekkehard Schlicht (Eds.)
Firms, Markets, and Contracts
1996. Corr. 2nd printing 1997.
ISBN 3-7908-0947-0

Thorsten Wichmann
Agricultural Technical Progress and the Development of a Dual Economy
1997. ISBN 3-7908-0960-8

Ulrich Woitek
Business Cycles
1997. ISBN 3-7908-0997-7

Michael Carlberg
International Economic Growth
1997. ISBN 3-7908-0995-0

Massimo Filippini
Elements of the Swiss Market for Electricity
1997. ISBN 3-7908-0996-9

Giuseppe Gaburro (Ed.)
Ethics and Economics
1997. ISBN 3-7908-0986-1

Frank Hoster/Heinz Welsch/Christoph Böhringer
CO_2 Abatement and Economic Structural Change in the European Internal Market
1997. ISBN 3-7908-1020-7

Christian M. Hafner
Nonlinear Time Series Analysis with Applications to Foreign Exchange Rate Volatility
1997. ISBN 3-7908-1041-X

Sardar M.N. Islam
Mathematical Economics of Multi-Level Optimisation
1998. ISBN 3-7908-1050-9

Sven-Morten Mentzel
Real Exchange Rate Movements
1998. ISBN 3-7908-1081-9

Lei Delsen/Eelke de Jong (Eds.)
The German and Dutch Economies
1998. ISBN 3-7908-1064-9

Mark Weder
Business Cycle Models with Indeterminacy
1998. ISBN 3-7908-1078-9

Tor Rødseth (Ed.)

Models for Multispecies Management

With 85 Figures
and 45 Tables

Physica-Verlag
A Springer-Verlag Company

Series Editors
Werner A. Müller
Martina Bihn

Editor
Prof. Tor Rødseth
Department of Economics
University of Bergen
Fosswinckelsgt. 6
N-5007 Bergen, Norway

ISBN-13:978-3-7908-1001-1 e-ISBN-13:978-3-642-99793-8
DOI: 10.1007/978-3-642-99793-8

Cataloging-in-Publication Data applied for
Die Deutsche Bibliothek – CIP-Einheitsaufnahme
Models for multispecies management: with 45 tables / Tor Rødseth (ed.). – Heidelberg; New York: Physica-Verl., 1998
(Contributions to economics)

softcover reprint of the hardcover 1st edition 1998

Softcover Design: Erich Kirchner, Heidelberg

SPIN 10567884 88/2202-5 4 3 2 1 0 – Printed on acid-free paper

Preface

The present book summarizes parts of the research made in the Norwegian Multispecies program. This program was initiated by the then Norwegian Council of Fisheries Research, which was later integrated in The Research Council of Norway.

Modelling has been a key element in the program. In the first place, modelling, be it singlespecies or multispecies modelling, is an efficient way of summing up existing knowledge, and performing analyses of different types on the basis of it. Furthermore, models may serve as a meeting ground between fishery researchers from different academic fields. This has been an important aspect of the Multispecies program, bringing together oceanographers, biologists, economists, mathematicians and statisticians in fruitful cooperation. That this has proved possible may be taken as an important result of the program. The multispecies aspect has perhaps underscored the need for inter-disciplinary cooperation.

The different authors are, of course, responsible for their respective contributions. Technical assistance in preparing the manuscript has been given by Knut Årrestad and Ian Knutsen, both at the University of Bergen.

Bergen, October 1997.

Tor Rødseth

Contents

Introduction

By Tor Rødseth.
Department of Economics, University of Bergen, 5020 Bergen, Norway.

Traditional Norwegian fisheries have shown great variations in catches and stock sizes over the years. The stock of Norwegian spring-spawning herring was almost extinguished by 1970. Consequently, this fishery was stopped until recent years. The stock of capelin in the Barents sea collapsed in 1985, and then, a new collapse occurred in 1993. Also, the stock of Norwegian arctic cod has shown great variations, with peaks in 1976, 1986 and 1993. Furthermore, stocks of birds preying on fish have deteriorated, while stocks of seal seem to have increased above the carrying capacity of the environment, causing extensive migration in search for food.

These changes in stocks and catches, of course, have had severe economic implications. To analyse the causes of these developments thus, has been given high priority by the authorities involved.

Traditionally, fisheries research has been studying each species in isolation. It is only until the last 15 years that multi-species approaches have been introduced into fisheries research, biological and economic. The recent dramatic changes in the Barents sea have given support to the suspicion that interactions between different species have played a great role in these developments. From ecological theory these interactions may take different forms, e.g. predator/prey relations, competition etc.

Biological interactions give rise to economic interactions. In addition to traditional relations between fishermen jointly harvesting a single stock, new relations are brought to the foreground by fishermen harvesting on different interacting stocks. Traditional problems of efficiency and distribution come into a new light when stock interactions are taken into consideration. These problems have both regional, national and international aspects.

The final aim of applied fisheries research, in principle, is to give to give managers tools which they can use to base their decisions on, and study the implications of decisions both with regard to efficiency and distributional issues. In view of the apparent great importance of multi-species interactions, new management tools based on multi-species approaches have to be developed, to supplement traditional single-species tools. This has been the general aim of the research program which is reported in the following articles.

The Institute of marine research in Bergen has had a multi-species research programme for several years, trying to model the ecosystem in the Barents sea, including the commercial most important fish stocks. Parallel to this work, bioeconomic research based on a multi-species approach has been going on at the Norwegian college of fishery science at the University of Tromsø.

In 1990 the Norwegian fisheries research council (now integrated into the Norwegian research council) established a national research program on multi-species management. The two research activities mentioned above have been in the core of the program. Besides this biological and economic research, other research groups have been included in the program. In the first place mathematical and statistical modelling at the Norwegian computing centre has added significantly to the program. Also work on political science aspects of multi-species management has been included.

The program formally was ended at the end of 1994. The purpose of this book is to report first of all the modelling work which has been carried out within the program. It may be added that much of the work in the multi-species program is carried further in a new national research program - Management of marine resources. This program was started in 1995 and will last for 5 years.

Modelling has been the core of the program. The work in this area can be divided into 4 elements:

- Ecological modelling
- Modelling of interacting fish stocks
- Bioeconomic modelling
- Management models.

The content of these expressions will become evident below.

ECOLOGICAL MODELLING

It is generally assumed that oceanographic and meteorological factors are the driving forces behind much of what is going on in the Barents sea. These factors work in many direct and indirect ways. Currents influence the spread of plankton into the area, which is the main food for several species. Sea temperature influences recruitment, growth and migratory patterns of several species and, consequently, also their interactions.

Under the somewhat imprecise heading above, research in this area has been carried out at the Institute of marine research in Bergen. In the research program what we here call ecological research has been treated as exogenous input to the models of interacting fish stocks.

The research under this heading is reported in the article by Giske, Skjoldal and Slagstad. At the bottom lies an oceanographic model. Wind, air temperature and tidal forces determine the currents and the circulation of water into and within the Barents sea. Ice cover, water temperature and salinity are simulated by the model and from this, primary production is simulated. The model produces projections of the inflow of plankton into the Barents sea and the distribution of plankton within the area. These projections can be compared with actual observations.

The output from the model, thus, is simulations of the supply of food to the stocks of fish and mammals in the area.

This builds a direct bridge to fishery biological models, either single or multi-species models.

Based on these results and behaviouristic assumptions the distribution of capelin in the Barents sea is predicted, and compared to actual observations. Since capelin plays a significant role in the Barents sea system, both as a prey and as a commercial stock in its own right, the possibility to make good predictions is of crucial importance. Moreover, this pioneering research on capelin may be extended to other stocks.

Generally, the prospect of linking together oceanographic and environmental models to models of fish populations seems very promising. More specifically, recruitment seems to be perhaps the most uncertain factor in population models of fish stocks. The traditionally assumed relation between spawning stock and recruitment seems to be influenced very much by factors which are the major elements in the research reported in this section, i.e. temperature, currents etc.

MODELS OF INTERACTING FISH STOCKS

The modelling work on fish stocks can be summarised under 4 headings:

- The MULTSPEC model
- The AGGMULT model
- The SYSTEM model
- The STOCKFISH model.

MULTSPEC

The MULTSPEC model has been the cornerstone of the research in fishery biology in the program, being a meeting ground between environmentalists, fishery biologists, economists and statisticians. The model is being developed at the Institute of marine research in Bergen (IMR). The model may be seen as a continuously ongoing project, summing up much of the institutes research. A survey of the model is given in the article by Tjelmeland and Bogstad.

The model is limited to the Barents sea, and to the following species:

- capelin
- herring
- cod
- harp seals
- minke whales.

In some cases the stocks are divided into sub-stocks. Furthermore, the stocks are divided into age groups, length groups and sexes. In the model there is also an area distribution.

At the basis for the modelling of fish stocks developments and interactions is oceanographic input which includes currents and sea temperature. These are used as exogenous input in the fish models. The current models determines both the drift of plankton and larvae, and thus, the food supply for plankton feeders. This, in turn, influences growth and migration of plankton feeders. Growth in length is generally taken as proportional to the feeding level. The stock development is also determined by fishing mortality and "natural mortality" the latter being a proxy for factors not included in the model and also for random elements.

Recruitment seems to be the most puzzling factor in fish populations dynamics. So far, only a simple relation between spawning stock size and recruitment has been implemented.

The fundamental element in models of this type, of course, is predation. Both competition and predator/prey relations are taken into account. Competition between plankton feeders is modelled by assuming that the total food supply is divided between stocks proportional to the stock biomass.

Most of the work on predation has been concentrated on the predation by cod, but recently, predation by mammals has been included in the model. The predation of one species on another is generally assumed to depend upon the number of individuals of the two species present. Cannibalism is also taken into account in the model.

It goes without saying that huge estimation problems are included in model building of this type. In addition to traditional estimation problems in single-species models, like estimation of recruitment and growth functions, one has to estimate effects of the two above-mentioned forms of interactions between species. As seen above, a simple theory to take care of competition is introduced in the model.

Much work has been done to estimate the effects of predation. The basic information on this is the analysis of stomach content. This type of information has been collected for many years by Norwegian and Russian marine researchers, to a large extent in co-operation.

The type of information involved is illustrated in the article by Tjelmeland and Bogstad, which gives diet composition for cod. Assessment of such information may be done in a theoretical framework. One such framework may be economic consumer theory, predicting that the composition of the diet would be determined by weighing marginal utility against marginal costs for the different sources of food. In our context, one possibility would be to evaluate both utility and costs in terms of energy. Energy costs may depend on the necessary travel distance to get a given type of feeding. Clearly, there would be some sort of substitution between preferred food far away, and not so interesting food nearby. Moreover, the article by Skjoldal and Giske analyses capelin demand for plankton on the basis of some sort of utility maximising.

The MULTSPEC model in a sense sums up much of the research going on at the Institute of marine research. Model runs may be made in different ways, under different assumptions and for different purposes. One may vary model runs by varying parameters and initial situations. Furthermore, one may also analyse the effects of uncertainty related to different elements in the model. Within the restrictions of the relations in the model, one may investigate the multi-species implications of keeping a given stock size at a certain level, or the consequences of a moratorium on a certain fishery.

A model of this type and of such dimension thus, may be considered as a general-purpose tool which can be used for a number of purposes. The output of the model is purely biological. This may be interesting in itself, but for management purposes, economic considerations have to be taken into account. The question naturally arises whether a big and detailed model like this is well suited as an input in economic analyses of management. This led to the construction of an aggregated version of MULTSPEC, described below.

AGGMULT

The AGGMULT model is, as the name indicate, an aggregated version of MULTSPEC. Initially, the purpose of this aggregated version was to establish a simpler model which could be used as an input in an economic multi-species model. It seems however, that such an aggregated model is interesting in its own right, and can also be used in purely biological contexts. One advantage with AGGMULT is that the information requirements are less than in MULTSPEC. It would be of great interest generally, to evaluate the "costs of aggregation" in terms of wrong predictions. So far this has not been done.

AGGMULT is described below in the article by Tjelmeland and Bogstad. The aggregation of MULTSPEC into AGGMULT is made by aggregating over fish length and area. The choice between the two models amounts to weighing manageability against detail. Parameters in AGGMULT are identical with those of MULTSPEC, and inputs and estimates in the former should be consistent with the latter model.

The AGGMULT model is the point of interface between the massive biological work at the Institute of marine research (and to some extent at the University of Tromsø), and the economic multi-species modelling taking place at the Norwegian college of fisheries at the University of Tromsø. This co-operation between biologists and economists may be considered one interesting aspect of the Multi-species program, and holds promising prospects for the future. And it is the model approach which brings the two areas of research together. Moreover, it seems that this co-operation has been professionally advantageous for both parties.

Quite a bit of work has gone into establishing compatibility between the two areas of modelling. This includes problems on a practical level concerning data compatibility and data transmission. Today is established an operational system making it possible to combine AGGMULT with the economic multi-species model ECONMULT, to which we shall return later.

THE SYSTEM MODEL

Is seems that migrations of herring have a significant influence on the ecosystem of the Barents sea. Depending on climatic and oceanographic conditions, herring migrates in and out between the Norwegian sea and the Barents sea. In the Barents sea, the predation of herring on capelin larvae influences the development of the capelin stock, and then, indirectly, the food supply to cod. This, in turn, may influence cannibalism among cod and also the food supply to marine mammals.

The collapse of the capelin stock in the middle of the 1980s and the beginning of the 1990s have been thought to be related to migration of herring into the Barents sea. These dramatic events, of course, have had great effects on the fisheries in the area. These experiences have led to the construction of the so-called SYSTEM MODEL for fisheries in the Norwegian sea and the Barents sea. This model is described in the article by Hamre and Hatlebakk.

In terms of aggregation, this model lies in between MULTSPEC and AGGMULT. It includes a flexible way of disaggregation into length groups. There is no area distribution, except that the herring stock is divided into two areas - the Norwegian sea and the Barents sea. This is made in order to try to grasp the influence of the herring stock on the ecology of the Barents sea.

Many elements in the SYSTEM MODEL corresponds to the similar elements in MULTSPEC. Recruitment depends on temperature and the size of the spawning stock. Characteristic temperature variations are used exogenously in the model. Growth is dependent on stock density and climatic factors. Predation is modelled by a predation matrix, which gives the probability of predation for different combinations of predator and prey. These probabilities depend on the differences in length between the two.

Model runs start from an initial stock situation. As is seen from the article, fairly good predictions of stock sized for cod, capelin and herring are produced.

The model was constructed with the specific purpose of explaining the dramatic changes in stock sizes in the Barents sea. The model has been calibrated to simulate the developments around 1985, and will be used to predict the latest collapse of the capelin stock. So far this has not been done.

THE STOCKFISH MODEL

Erling Moxnes' article describes an aggregate multi-species model which is constructed for the purpose of policy analysis under uncertainty. It has been an objective to make the model compatible with AGGMULT, described above, and ECONMULT, to be described below. Furthermore, an objective has been to make the model as simple as possible, making it suitable to produce the numerous model runs which are involved in a stochastic policy analysis.

Much of the biological input and data stems from co-operation with the Institute of marine research. The model comprises the stocks of capelin, herring and cod.

The article illustrates the possibility of constructing a simple model on an acceptable empirical basis. Based on data on stomach content as left-side variables, estimates of the influence of the density of prey and the biomass of predators are made with acceptable significance. Also, relations estimating the growth of plankton, recruitment of herring, capelin and cod are estimated, with, in some cases surprisingly high level of significance.

On the bases of these estimated relations, sizes of the three stocks are estimated and compared to observed stock developments. Except for capelin the model predictions seem to be fairly good.

ECONOMIC MODELLING

The cornerstone in the economic modelling in the multi-species program is the ECONMULT model, developed by the Norwegian college of fishery science at the University of Tromsø. The model is described in the article by Eide and Flaaten. It is constructed with the purpose of being integrated with a biological model. That is, the biological side of the interactions between economic and biological factors are taken care of by the biological modelling. Moreover, the ECONMULT is a rather general economic model, being compatible with different biological models. So far, the integration with AGGMULT has been the basic element of the work.

Variables in the model relate to stocks, cohorts and vessel groups. The former two, of course, must correspond to the similar variables in the biological model. Vessel groups are differentiated according to size and gear type.

Production (catches) in each fleet group is assumed to depend on effort (number of vessels times number of fishing days) and stock sizes (biomass). A Cobb-Douglas production function is used. Fixed and variable costs are included. As for the market side, prices are either taken as fixed, or linearly dependent upon catch.

The model can be used in a number of ways and under different sets of assumptions. The basic approach is simulations, but these can be done in different directions.

Simulations based on an earlier, simplified version of the biology (the MULTSIMP model) included cod and capelin. Nine simulations were made, based on different combinations of effort in the two fisheries, giving present value of the resource rent for a 27 years period. Thus, such simulations can be used to get indications of what is an optimal policy. Furthermore, simulations were made to assess the increase in profits by using only the most cost-effective vessel groups.

Based on the latest version of AGGMULT, a number of simulations were made, which are reviewed in the article by Eide and Flaaten. These are made under different assumptions on quotas, effort, costs, contribution margins etc.

ECONMULT can be considered as a general-purpose model, which can be used under many different assumptions and combined with different biological versions. The advantage of this is the rich variety of applications. The model can be used as a simulations model, producing scenarios under different assumptions. It can be used as a decision model, estimating the consequences of management decisions, be they firms in the industry or government agencies. And, finally, as indicated above, it can be used as an optimisation model, finding the optimal mix of efforts in the different interacting fisheries.

Many further developments and uses of ECONMULT are possible. The uses of the model have so far not included the important predation by marine mammals, which should be included in further applications. This also applies for shrimps which have important interactions with cod.

The differentiation into vessel groups forms a basis for using the model to analyse problems of distribution. This also includes problems of regional distribution, which is an important issue in Norwegian fisheries.

MANAGEMENT

Biological and economic models like the ones reviewed above give a basis for management decisions by the authorities in fishery policy. The essential lesson from fishery economics is that an unregulated fishery in general produces non-optimal results. However, it may be said that neither the results from biological, nor from economic analysis, adequately simulate the concrete decision problems in fishery management. Traditionally, biological advise has been little concerned with the vast uncertainties involved in the research area, perhaps because managers like point estimates better than probability distributions. And economic research so far, has been little concerned with the many restrictions - political and other - which managers feel they have to take into account in their fishery policies decisions.

Thus, there seems to be a missing link between the output of biological and economic research, on the one side, and management decision making on the other. Considerable work seem to be necessary in order to build the bridge between these elements.

An illustration of an approach in this direction is reviewed in the article by Hagen, Hatlebakk and Schweder. The work within in the Scenario Barents Sea project is somewhat inspired by similar work within the scientific committee of the International whaling commission (IWC).

A scenario model is defined as consisting of three elements:

- A simulation model
- A management procedure
- A performance measure.

The simulation model has two ingredients: a simulation of the ecosystem of the Barents sea, including the main biological and environmental factors, and human activities including research surveys and fishing activity. Biological factors are the usual ones in multi-species fishery models, and this part of the work is based on the research going on at the Institute of marine research. The stocks included are cod, capelin and herring.

The human activity includes elements which are not usually modelled in this area.

Besides fishing activity, research surveys is modelled, giving data input. Moreover, overfishing and discarding may be included. Uncertainty in survey data is taken into account.

The main element in the management procedure is a quota routine fixing a quota on the basis of estimated stock sizes. Stock sizes are estimated on the basis of catches and research surveys. The model is sequential, in the sense that the fixed quotas are input to the simulation model for the next period.

Management strategies are based on the relation between number of recruits and spawning stock biomass. A median, a high and a low alternative are implemented, giving a corresponding quota.

Performance measures are needed in order to evaluate different strategies. At the stage of the research presented in the article, limited experiments with response variables have been made. In the first place response variables are mainly limited to cod, being the commercially most interesting stock. Moreover, economic response variables have so far not been implemented. Thus, rather few and mainly biological, of a large number of possible biological and economic performance measures have so far been tested out. Examples of measures which have been tested out are mean yearly catches, minimum and maximum catches and minimum stock biomass.

Perhaps the most promising aspect of this type of research lies in its generality. Updated biological knowledge may be combined with many different management strategies giving a number of possible performance measures. The latter can be used to address the fact that fishery policies must be based on a multi-dimensional preference indicator.

Ecological Modelling for Fisheries

By Jarl Giske[1], Hein Rune Skjoldal[2] and Dag Slagstad[3].
[1]Department of Fisheries and Marine Biology, University of Bergen, N-5020 Bergen, Norway.
[2]Institute of Marine Research, Postboks 1870 Nordnes, N-5024 Bergen, Norway.
[3]SINTEF Electronics and Cybernetics Automatic Control, N-7034 Trondheim NTH, Norway.

HARVESTING A DYNAMIC ECOSYSTEM

Historical perspective

Large fluctuations in the fisheries have been a characteristic feature of great importance in the history of the Norwegian people. Herring periods have been times with large fisheries for herring spawning along the west coast of Norway, alternating with periods when the herring was gone. Herring periods and periods without herring appear to have been of cyclic nature with a periodicity of about a century (Devold 1963, Dragesund & al. 1980, Skjoldal 1990, Øiestad 1990). Towards the end of the last century, a common explanation for such fluctuations was that it reflected variable migration routes and therefore variable availability of perceived stable fish stocks to coastal fisheries. When ICES was established in 1902, a standing committee called the Migration Committee was created to deal with such issues (Sinclair & Solemdal 1988). However, in his classical book, Johan Hjort (1914) identified large fluctuations in stock size due to variable recruitment as a major source for fluctuations in the fisheries. Variable recruitment is an inherent property related to the reproductive mode of fishes with high fecundity and which produce a high number of larvae that are part of the planktonic system (Skjoldal & Melle 1989).

The large fluctuations in the fish stocks have caused many fishery crises that have focused the attention of the public, managers, and politicians on the need for better knowledge of the underlying causes. In the 1870s when herring had disappeared, G. O. Sars and H. Mohn were given the task to organise the "Vøringen" expeditions which explored the Norwegian Sea with the main motivation to search for the lost herring. Around the turn of the century there was a crisis in the cod fishery in northern Norway which coincided with massive seal invasions (Skjoldal 1990). Species interactions were considered to play an important part in current explanations for the crisis at that time. Fishermen believed that killer whales had large baleen whales and seals as their main prey. With the depletion of the stock of large whales by the whaling industry developed by Sven Foyn from the late 1860s, killer whales were believed to no longer frequent the area which then was open for seals to invade (Hjort 1902).

The events at the turn of the century took place during a cold climatic period. 1902 was a cold year with extensive ice cover during the winter. 1903 marked the peak in the fishery crisis with low catches in the cod fishery with small cod in poor condition, massive seal invasion during winter, and large number of dead seabirds reported. From 1902 to 1904 there was a warming trend and in 1904 strong year classes of cod and herring were formed (Hjort 1914, Skjoldal 1990). The fishery crisis resulted in increased research effort. With the new research vessel "Michael Sars" built in 1900, systematic oceanographic and fishery investigations were carried out that resulted in benchmark publications (Helland-Hansen & Nansen 1909, Hjort 1914) during the "golden age" in Norwegian oceanographic research that ended during the first World war.

Nearly a century later, during the 1980s, events that resembled the events at the turn of the century, again took place in the Barents Sea. The capelin stock collapsed, the cod stock decreased and cod were small and in poor condition, there were seal invasions along the coasts of northern Norway, and there was heavy mortality in seabird colonies at Spitsbergen and in northern Norway (Skjoldal 1990, Blindheim & Skjoldal 1993). Due to an increasing awareness of the need to understand and take into account ecological relationships as a basis for management, the research program PRO MARE (Sakshaug & al. 1991, 1994) was run from 1984 to 1989 during the period of dramatic changes in the Barents Sea ecosystem. Along with results from routine fishery investigations, PRO MARE has provided us with a reasonable understanding of the main ecological features and causes for variability in the Barents Sea ecosystem (Skjoldal & Rey 1989, Sakshaug & Skjoldal 1989, Skjoldal 1990, Sakshaug & al. 1991, 1994).

The ecological events and the fishery crisis in the Barents Sea during the 1980s followed the longest and coldest period in the Barents Sea since the cold period at the turn of the century. The cold period from 1977 to 1981 coincided with the arrival of the "great salinity anomaly" which has been related to a large outflow of water from the Arctic Ocean through the East Greenland Current in the 1960s (Dickson & al. 1988). This anomaly affected or coincided with changes in fish stocks along its route of progress in the Nordic and adjoining seas (Blindheim & Skjoldal 1993). Bridging the two cold periods at the turn of the century and in the 1970/80s was a long-term climatic oscillation with a warm climate mode from about 1920 to the 1960s (Loeng & al. 1992) which also constituted the main part of the last herring period (Blindheim & Skjoldal 1993). The Barents Sea cod stock has also shown large changes over this time period from a small stock size dominated by first time spawners at the turn of the century, to a large stock size with multiple year classes in the spawning stock in the warm period, to the state in the late 1980s with again a small stock size with a young age composition (Hjort 1914, Jørgensen 1988, Godø & Skjoldal 1991). The lesson to be learned from a retrospective analysis of the history is that fish stocks show large changes in response to oscillations in the ocean climate. The challenge to fishery managers has been and will always be to regulate the fishing pressure on stocks against the background of large natural fluctuations (Skjoldal 1990).

Methods for monitoring and assessing the state of fish stocks have improved markedly over the last decades. Using echo sounders and trawls during scientific surveys as well as information from commercial catches, data are provided on size and age composition of stocks. This provides a basis for assessing the likely development of stocks and for setting recommended catch quotas. Fisheries management has up to now been mainly single-species oriented. However, the need for taking species interactions into account has been recognised by fishery managers, and in 1990 a 5 years research program on Multispecies management was started in Norway with focus on the Barents Sea ecosystem. Major emphasis has been given to the development of a multispecies model, MULTSPEC, with capelin, cod and herring as key species (Bogstad & al. 1992, Tjelmeland & Bogstad 1997 (this book), Ulltang 1995). Environmental variability has also been considered in the Multispecies management program and development of ecological models has started. The aim has been to develop ecological models where the dynamics of main components of the ecosystem are linked to the physical driving forces and biological interactions which govern the variability of the ecosystem. Capelin is an important intermediate link in the pelagic food web in the Barents Sea, and it was chosen as the focal species for the initial development of ecological models (Giske et al. 1992a,b).

The Barents Sea

The Barents Sea is a high latitude ecosystem located between 70 and 80°N. The main features of the physical and biological conditions of the ecosystem are presented in several recent reviews (Dragesund & Gjøsæter 1988; Loeng 1989a,b; Sakshaug & Skjoldal 1989; Skjoldal & Rey 1989, Sakshaug & al. 1991, 1992).

The water circulation is characterised by an inflow of relatively warm Atlantic water and coastal water from the west. The Atlantic water is separated from the cold Arctic water in the north by a sharp polar front in the western Barents Sea (Fig. 1) which is topographically determined. In the eastern Barents Sea there is extensive mixing and transformation of the inflowing water into Arctic water, and the polar front is less sharp and more variable in location (Loeng 1989a; Midttun 1989). There is large interannual variability in ocean climate related to variable strength of the Atlantic water inflow and exchange of cold Arctic water (Midttun 1985, 1989; Loeng 1989a,b; Ådlandsvik & Loeng 1991). This variability is cyclic with a dominant cycle length of about four years (Loeng & al. 1992). The year class strength of commercially and ecologically important fish species has been empirically related to these cyclic ocean climate changes (Helland-Hansen & Nansen 1909; Sætersdal & Loeng 1987; Loeng 1989b). Through trophic interactions, variable recruitment of one species affects populations of other species in the ecosystem (Skjoldal & Rey 1989; Hamre 1991). The variable ocean climate has also a direct effect on the lower trophic levels through the influence of ice on primary production. An ice edge phytoplankton bloom is generated when ice melts. This bloom sweeps across the sea as a band of high production as the ice retreats northwards during summer (Sakshaug & Skjoldal 1989). The ice edge bloom triggers reproduction and nourishes the new generation of zooplankton. The spring bloom in Atlantic water, not influenced by sea ice, is in contrast developing more slowly and is protracted in response to the seasonal thermocline formation (Rey & al. 1987; Skjoldal & al. 1987; Skjoldal & Rey 1989).

The transformation of inflowing Atlantic water into Arctic water makes the Barents

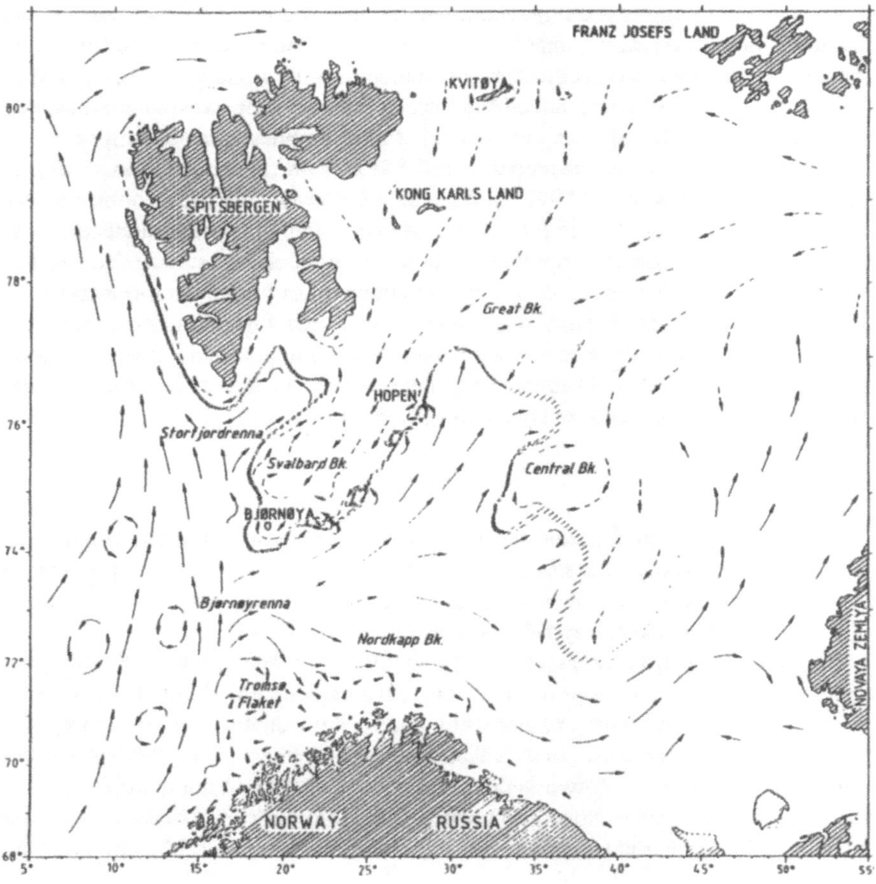

Figure 1. The dominant surface current system of the Barents Sea. Solid arrows: Atlantic currents; broken arrows: Arctic currents; dotted arrows: coastal currents. The polar front is indicated with a hatched line. Redrawn by Giske & al. (1992a) from Loeng (1989b).

Sea into a zoogeographical transition zone and into an ecosystem strongly dependent on plankton transport with incoming currents. The copepod *Calanus finmarchicus*, which is a key herbivore, performs a marked seasonal vertical migration in the Norwegian Sea (Østvedt 1955). During winter *C. finmarchicus* resides below the threshold depth at the entrance to the Barents Sea. Horizontal advective transport and seasonal vertical migration interact to produce large variability in standing stock of zooplankton in the western and central Barents Sea (Skjoldal & Rey 1989). Major inflows during winter will transport warmer Atlantic water but not copepods, while inflows during summer also will bring in food for the pelagic fishes (see section Inflow of *C. finmarchicus* from Norwegian Sea).

The food web of the Barents Sea ecosystem is relatively simple with a few domi-

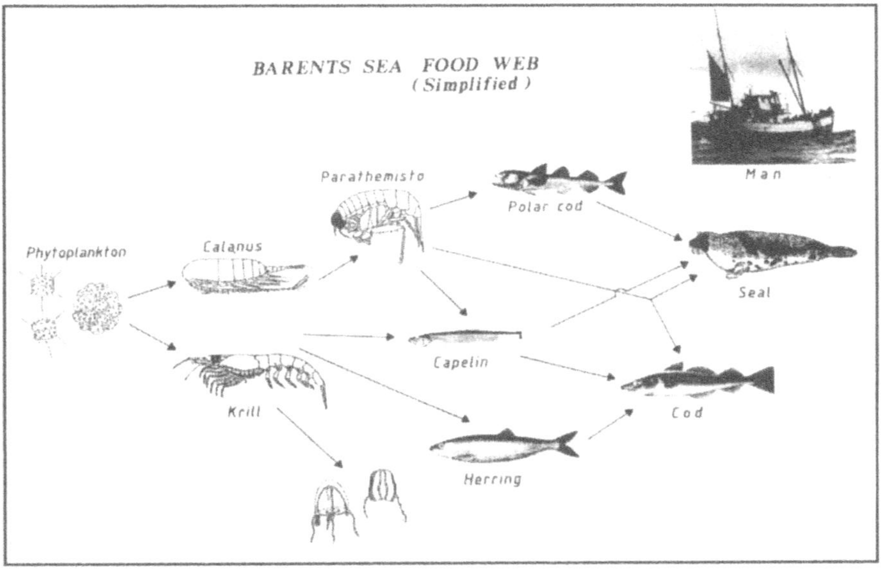

Figure 2. Simplified food web of the Barents Sea.

nant species at each trophic level (Fig. 2). Diatoms and the colony-forming flagellate *Phaeocystis pouchetii* are dominant forms of phytoplankton during the annual spring bloom. Calanoid copepods (*Calanus finmarchicus* and *C. glacialis*) and krill (*Thysanoessa* spp.) are the dominant herbivores. Capelin (*Mallotus villosus*), herring (*Clupea harengus*) and polar cod (*Boreogadus saida*) are dominant planktivorous fish, while cod (*Gadus morhua*), sea mammals and seabirds are important piscivores (Dragesund & Gjøsæter 1988; Loeng 1989a).

Amphipods (*Parathemisto* spp.) and gelatinous zooplankton (medusae and ctenophores) are important invertebrate predators.

Capelin

Capelin is of particular importance as an intermediate link in the food web. This species is a small salmonid fish with a short life span; the majority of individuals spawn only once (Hamre 1985, 1991, Vilhjálmsson 1994). The immature capelin performs a large scale feeding migration northwards in the Barents Sea during summer (Fig. 3). This migration can be viewed as an adaptation, whereby the fish follows the retreating ice edge with a time delay to exploit the ice edge phytoplankton production (Sakshaug & Skjoldal 1989; Skjoldal & Rey 1989). A large capelin stock needs a large feeding area to sustain its production, and the seasonal feeding migration is in this respect a necessary requirement (Skjoldal & Rey 1989).

The adult stage of capelin is characterised by reaching a minimum size rather than a minimum age (Forberg & Tjelmeland 1985; Hamre 1985, 1991). Growth is seasonal with large inter-annual variability in rate (Gjøsæter 1985; Gjøsæter & Loeng 1987;

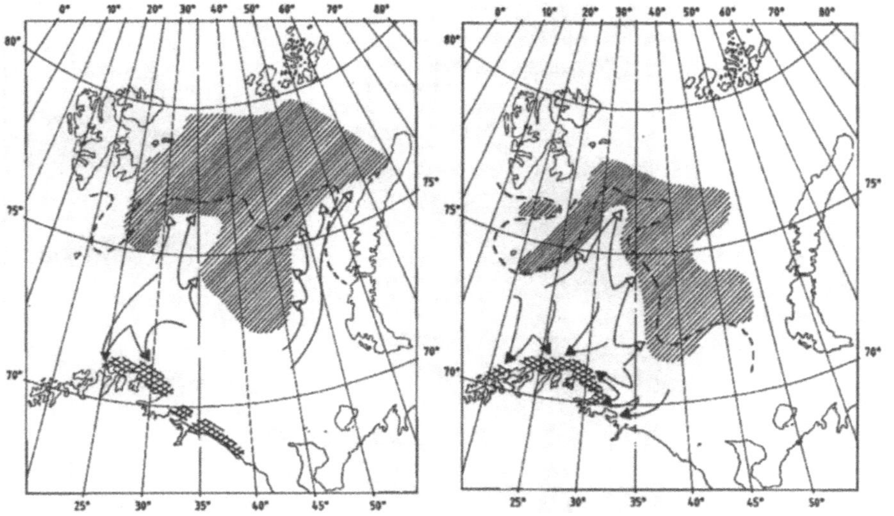

Figure 3. Feeding distributions (hatched) and spawning grounds (double hatched) of Barents Sea capelin in warm (above) and cold (below) years. White arrows indicate feeding migrations, black arrows spawning migrations. Redrawn by Giske & al. (1992a) from Ozhigin & Luka (1985) and Loeng (1989b).

Loeng 1989b). Individuals which have reached a size of 14-15 cm in the autumn will mature and be part of the spawning stock which migrates to the coasts of northern Norway and Murman to spawn in late winter. The immatures overwinter in the polar front region from where they migrate northwards during the following summer (Fig. 3).

There are many spawning sites along the whole coastal stretch from Troms county to the Kola Peninsula (Fig. 3). The climatic fluctuations influence the spawning migration and choice of spawning sites, with trends towards a shift from westerly spawning grounds in cold periods to easterly spawning grounds in warm periods (Fig. 3; Ozhigin & Ushakov 1985; Ushakov & Ozhigin 1987). Following a relatively long incubation period in the sediments, the hatched larvae are released into the water in early summer (Alvheim 1985). From here they drift with the currents eastwards and northwards to the nursery and feeding grounds of the central Barents Sea. From easterly spawning grounds in warm periods larval transport tend to be towards the eastern Barents Sea, with a time-delayed shift towards more westerly distributions in cold periods (Ozhigin & Luka 1985; Ushakov & Ozhigin 1987; Loeng 1989b).

The geographic closure of the capelin life cycle is relatively loose in the sense that larvae drift from widely distributed spawning grounds along the coast of the southern Barents Sea to wide target areas in the central Barents Sea. Loss of recruits through vagrancy (Sinclair 1988) is therefore less likely for capelin than for species with more complex drift routes, such as the Norwegian spring spawning herring and the

Norwegian arctic cod.

Cod is the major predator on capelin, with a potentially high impact on the dynamics of capelin. Cod is mainly distributed in the southwestern (warmer) part of the Barents Sea, but young cod are known to migrate as far north as the Central Bank (Nakken & Raknes 1987). The spatial overlap between capelin and its major predator is thus largest in the spawning period, when capelin must pass this high risk area to come to the spawning grounds. In this period cod feeds almost exclusively on capelin.

Predation from juvenile herring appears to be an important factor for the recruitment success of capelin (Skjoldal & Rey 1989; Hamre 1991; Fossum 1992). Juvenile herring from strong year classes have their main nursery area in the southern Barents Sea. They are thus distributed in the area through which capelin larvae drift from the coastal spawning grounds to their nursery area in the central Barents Sea.

STATE-OF-THE-ART IN ECOSYSTEM MODELLING

Complexity and management

Complexity

The complexity of nature, and even of a relatively simple subset of it, is so overwhelming that it is impossible to describe it in detail, neither by words nor models. Our concepts are simplifications, and must be so to reveal understanding. However, this simplification has implications for predicting the future, as many processes and variables cannot be accounted for in verbal or mathematical models. And extensive inclusion of processes and variables need not improve the predictive power of a simulation model, as each new process and parameter also adds some uncertainty (Håkanson 1995). The benefit of mathematical modelling relative to a subjective judgement is that the mathematical language allows us to express current knowledge and assumptions in a precise form, and to compare the effects of forces and processes that may act on widely separated spatial and temporal scales. However, with the limitation of knowledge, the output from any model should be evaluated by the subjective judgement of an expert.

Current fisheries management models

Fisheries management in the Northeast Atlantic has for a long time been based on a combination of field investigations and modelling. The core of the common modelling concept, Virtual Population Analysis (VPA), is a cohort analysis of population dynamics. VPA models are based on catch data to calculate mortality rate of the fish population (Gulland 1965; Pope 1972, 1979). Population size and size frequencies are estimated from a data set on age composition, growth, fishing effort and recruitment. This technique is further developed to multispecies models (Magnússon 1995), where interactions between stocks (predation) are estimated from stomach analyses. VPA has proven to be a good tool for fisheries management. However, VPA has two limitations: 1) it assumes constant environment, and 2) since the pre-recruitment period in the life of the fish is not included, VPA is not suited for fish that lives only for a short time after recruitment. In cases where the physical environment varies extensively over short intervals, and hence growth and survival prospects of the commercial fishes as well as of their prey (and predators), an approach where environmental variability is accounted for is needed. This is the situation in the Barents Sea. In addition, the planktivore in the Barents Sea, the capelin *Mallotus villosus*, lives only for 1-3 years after recruitment. For both these reasons the Institute of Marine Research has been developing a quite different set of models for the major fish species in the Barents Sea. The central tool, a multispecies fisheries model (MULTSPEC, see next chapter) includes predator prey relationships as well as food and temperature limitation in growth.

Since predictability of VPA based models relies on a statistical analysis of the years past, such models are far more capable of dealing with stable systems than with fluctuations and perturbations. Therefore, model predictions have not been good in years when stock sizes or environment have deviated from the normal. This has led to an interest for studying fish stock dynamics by means of other modelling concepts. Two properties are desirable: 1) feeding and growth to be dynamically coupled to food concentration (production and advection) and 2) migration and predator-prey relationships to be described in theoretical rather than empirical terms.

While a wide range of empirically tuned functions may produce fits with observations in stable systems, representation of the forcing functions and their influence in processes are crucial for the performance of theoretical models. Generally, a theoretical formulation of a process will demand a higher number of variables, constants and parameter values, many of which may not have been measured. This will probably lead to lower goodness-of-fit of any new theoretical model compared to existing empirical models. Quantification of the theoretically derived constants and variables is necessary for a theoretical model to reach the same level of goodness-of-fit as an empirical model, even in systems of some instability. The strength of theoretical models, however, is to handle unexperienced situations, which is when modelling is most valued.

Balchen and colleagues were the first in Norway to use theoretical models in studies of fish in ecosystems (Slagstad & al. 1975; Balchen 1976a,b). Not only have several new generations of computers passed since then, quantitative ecological theory has also evolved. Today, there are a number of dynamic models available covering water transportation, primary production and zooplankton (see Aksnes & Lie 1990, Aksnes & al. 1995), while fewer dynamic models of fish exist. The theoretical constraints has been the formal representations of mortality and of the biological forces that motivate the behaviour of the individuals. During recent years, spatial distributions have been studied by two methods: life history theory (Werner & Gilliam 1984; Aksnes & Giske 1990) and dynamic optimisation (Gilliam 1982; Clark & Levy 1988; Houston & al. 1988; Mangel & Clark 1988). Both rely on the assumption that fish behaviour can be understood from evolutionary forces. A prerequisite for applying them in natural habitats is that natural mortality rates can be calculated. This can be done by the method of Aksnes & Giske (1993) and Giske & al. (1994).

Goals of ecological modelling in the Multispecies Program

Even in a high-latitude area as the Barents Sea, the overwhelming complexity of the ecosystem demands a focus on a few central factors in the modelling (Håkanson 1995). Therefore, and as the Barents Sea is an open area heavily influenced of the inflow of warm Atlantic water and zooplankton from the Norwegian Sea (Midttun 1985, 1989; Loeng 1989a,b; Ådlandsvik & Loeng 1991, Loeng & al. 1992, Sakshaug & al. 1991, 1992), modelling advection of zooplankton has been seen as a central task.

However, while several research groups have had considerable success in modelling the dynamics of nutrients and primary production (e.g. Aksnes & Lie 1990, Aksnes & al. 1995, Baretta & al. 1995), success in modelling higher trophic levels is still lacking, mainly due to the representation of the spatial processes. Successful representation of spatial dynamics of zooplankton and fish have been hindered by ignorance of, or low emphasis on, the role of advective transport (e.g. Giske & al. 1991 and Salvanes & al. 1992, vs. Salvanes & al. 1995), but often even more of the lack of understanding of how the internal motivation of the individual fish and zooplankter influence their behaviour and hence the dynamics of their population and the entire ecosystem.

Both for the force of advection and motivation, there is a big difference between phytoplankton and zooplankton (and fish). Phytoplankton drift passively with the currents, and are as such completely controlled by physical forces. However, their rate of multiplication under favourable local conditions may enable them to reproduce much faster than they are advected. Zooplankton, on the other hand, have generation times of weeks to years, and may not by reproduction counteract moderate currents. This was evident in the western Norwegian fjord Masfjorden (Aksnes & al. 1989, Giske & al. 1991), where the biomass of phytoplankton could be modelled from the local conditions, while understanding zooplankton dynamics required knowledge of the water transportation. Both this small fjord and the Barents Sea are open systems connected to the larger Norwegian Sea Large Marine Ecosystem (LME) (Blindheim & Skjoldal 1993, Sherman & al. 1993). But while phytoplankton are passively advected, zooplankton have the locomotory capacity to perform extensive vertical migrations on several time scales. Fish may even swim in opposite direction to currents, and hydrodynamics are poor indicators of horizontal distribution of all but the youngest individuals.

A major focus in this program has therefore been to model the motivation for behaviour of individuals. This has led us into complex models, with the hope one day to be able to synthesise the impact of these evolutionary and ecological forces into simpler expressions in later generations of ecosystem models. In the last part of this chapter we will demonstrate an attempt to model the distribution, growth and survival of capelin, as determined by natural selection (reproduction), ecological forces (food and predators) and the physical setting (hydrodynamics, temperature, bottom topography).

Hydrodynamics

The Navier Stokes equation (Box 1) represents an excellent model of the hydrodynamical process in general. This equation is also widely used to calculate any flow of liquid or gas and in models for weather forecasting. In contrast to biological processes, the model for hydrodynamical processes are well known. The main problem is how to simplify and represent the equation in a computer.

The forces that govern the circulation pattern may be different from one sea to another and will also depend on the season. In has been shown that wind, as in most other seas, is an important driving force for the circulation and inflow of Atlantic water into the Barents Sea (Ådlandsvik & Loeng 1991). Slagstad (1987) showed that it is necessary to include density variations in the models in order to obtain flow fields similar to what is observed. Density variations in the Barents Sea are caused by inflow of low saline water long the coast of Finnmark (continuation of the Norwegian Coastal current) and melting/freezing of ice in the northern Barents Sea. Since the Barents Sea has boundaries to other large Seas (Norwegian, Arctic and the Kara Seas), processes in these nearby areas also have a pronounced effect on the Barents Sea circulation pattern.

Baroclinic and barotropic currents

The pressure at a certain depth depends on the weight of the water column above and the atmospherical pressure. This weight is again a function of the water density. As the density depends on the temperature and salinity, the pressure will vary accordingly. An imaginary surface that connects all the points having the same pressure is called an isobaric surface. If temperature and salinity are not variable in space, this isobaric surface is parallel to the ocean surface.

In the same way one could make an imaginary surface by connecting all the points having the same density, isopycnal surfaces. When the isobaric and isopycnal surfaces are parallel to each other, we call this a barotropic field of mass. If the density varies horizontally, the isobaric and isopycnal surfaces may be inclined to each other and the mass field is baroclinic. The flow field driven by a baroclinic mass field as called baroclinic flow. The Barents Sea is characterised by large horizontal density gradients, which mean the baroclinic flow is important for this area.

When the flow is barotropic, the velocity will be the same in the whole water column, whereas one in the baroclinic case have currents which has a variable magnitude and direction with depth. Shelf seas at high latitude are often well mixed during the winter and spring. Many, so-called depth integrated circulation models, have been developed for this situation. It is only the central part of the Barents Sea, in the core of the inflowing Atlantic water, that can be regarded as well mixed during the winter and depth integrated models would not simulate the flow field with great realism. A hydrodynamical model of the Barents Sea must, therefore, be of baroclinic type having a vertical structure (3-dimensional).

There are several kind of models that may be used for to simulate the baroclinic flow field of the Barents Sea. The most used models are of the so-called level type which mean that the vertical water column is divided in layers (or levels) each with a fixed thickness, independent of time. The level models may be divided into categories depending on how the vertical levels are divided. The σ-coordinate models divide the water column into a fixed number of levels, each occupying the same fraction of the water column. If a certain level represent 1 % of the water column, the depth of this level will be 1 m if the depth is 100 m and 20 m if the depth is 2000 m. The main reason for using σ-coordinates is the ability to resolve some of the processes near the bottom which are important for the energy dissipation.

In the z-coordinate models the thickness of the levels is independent of the water depth except for the level that happens to be near the bottom which will change in thickness. However, the surface levels will have a fixed thickness, independent of the water depth.

Most models available today seem to handle the numerical in an acceptable way. Although improvements can be done, great progresses can be achieved through a finer horizontal grid. Usually, grid point distance of 20 km has been used for the Barents Sea, but many important processes have a length scale which cannot be resolved by this grid. The lower limits of horizontal grid point distance is determined by the computer capacity (memory and computational speed). If the horizontal grid point distance is reduced by a factor of 2, the required memory capacity will increase by a factor of 4 and computational work by a factor of 8.

There are several models that have the potential to be usable for the Barents Sea, but the model should also be able to produce and melt ice in response to atmospherical forcing and water temperature. Most of the hydrodynamical simulations shown and used as input for biological simulations in this book are taken from a level model using z-coordinates.

Hydrodynamical model

In the ocean there are several forces connected to the movement of water masses. We may divide these forces into primary forces which produce water movements and secondary forces which are a result of water movements. The primary forces are horizontal pressure caused by variations in the density, sea surface elevation or air pressure, wind and tidal forces caused by variation in gravity forces from the moon and the sun. The secondary forces are the Coriolis force (which is due the rotation of the Earth and acts on moving water with a force which is to the right of the direction of motion on the Northern hemisphere and to the left on the Southern hemisphere) and friction. Most models of ocean currents are dynamic (prognostic) which means that the forces acting on a certain water mass lead to acceleration according to Newton's 2. law of motion:

$$a = \frac{f}{m} \tag{1}$$

where a is the acceleration and f is the net force acting on the mass m. Integration in time of acceleration gives the velocity. Based on this equation and the equation of continuity (which state that water masses should be conserved) a set of partial differential equations can be developed.

The model described (Box 1) is a 3-dimensional, baroclinic, finite-difference "level model" that is defined by a sequence of fixed but permeable levels. Each level has a fixed thickness, except the level near the surface and the level that happens to be near the bottom. Thus, the number of levels will be a function of horizontal coordinates.

Box 1: The equations of the hydrodynamical model

The equations describing the motion of incompressible water on a rotating earth, incorporating horizontal and vertical diffusion of momentum, can be written:

Accelleration in x-direction:

$$\frac{\partial u}{\partial t} = fv - u\frac{\partial u}{\partial x} - v\frac{\partial u}{\partial y} - w\frac{\partial u}{\partial z} - \frac{1}{\rho}\frac{\partial p}{\partial x} + A_v \nabla^2 u + \frac{\partial}{\partial z} A_v \frac{\partial u}{\partial z} \tag{2}$$

Accelleration in y-direction:

$$\frac{\partial v}{\partial t} = fuv - u\frac{\partial v}{\partial x} - v\frac{\partial v}{\partial y} - w\frac{\partial v}{\partial z} - \frac{1}{\rho}\frac{\partial p}{\partial y} + A_h \nabla^2 v + \frac{\partial}{\partial z} A_v \frac{\partial v}{\partial z} \tag{3}$$

Vertical velocity is found from the equation of continuity:

$$\frac{\partial w}{\partial z} + \frac{\partial u}{\partial x} + \frac{\partial v}{\partial y} = 0 \tag{4}$$

and surface elevation

$$h = \int w_1 dt \tag{5}$$

where

u, v	- horizontal velocity components in x- and y-direction, respectively
w	- vertical velocity component
w_1	- vertical velocity of surface elevation, i.e. upper layer
h	- height of the free surface from the undisturbed mean
f	- Coriolis parameter
ρ	- density
A_h	- horizontal eddy diffusion of momentum
A_v	- vertical eddy diffusion of momentum
p	- pressure found by the hydrostatic equation

$$p = \int_z^h rgdz + Pa \tag{6}$$

where *Pa* is the atmospheric pressure.

The space-time variations of water density ρ are found by the functional relationship

$$\rho = \rho(S,T) \tag{7}$$

where *S* and *T* are the salinity and temperature of the water, respectively. These scalar fields can be modelled by a balance equation of the form

$$\frac{\partial c}{\partial t} = \frac{\partial}{\partial x}(uc) - \frac{\partial}{\partial y}(vc) - \frac{\partial}{\partial z}(wc) + K_h \nabla^2 c + \frac{\partial}{\partial z}\left(K_v \frac{\partial c}{\partial z}\right) + \delta_c \tag{8}$$

where

- c - S or T
- K_h - horizontal eddy diffusion of salinity and temperature
- K_v - vertical eddy diffusion of salinity and temperature
- δ_c - thermodynamic interaction between the upper layer and the atmosphere (c = T) or supply of salt and fresh water during freezing of melting, respectively (c = S) and the operator ∇^{-2} means

$$\nabla^2 \Theta = \frac{\partial^2 \Theta}{\partial x^2} + \frac{\partial^2 \Theta}{\partial y^2} \tag{9}$$

where Θ is any function of x and y.

Vertical mixing is calculated as a function of the Richardsons number, wave height and parametrisation of tidal mixing in shallow areas.

Phytoplankton dynamics

The phytoplankton model is based on nitrogen as the limiting nutrient and have two state variables: concentration of phytoplankton and nitrogen. The phytoplankton growth rate is a function of irradiance and the concentration of nitrogen. Loss of phytoplankton is due to respiration, sedimentation, and grazing from zooplankton. Model equations are given in Box 2. A recent review of phytoplankton modelling is given in Aksnes & al. (1995).

Zooplankton dynamics

In the Barents Sea there are two dominating species of copepods which serve as food for fish larvae and planktivorous fish. In the Atlantic water (south of the Polar Front) we find *Calanus finmarchicus* whereas *C. glacialis* is found in water masses of Arctic origin. *C. finmarchicus* is regarded to have its core distribution area in the Norwegian Sea. Spawning starts in spring as soon as there are enough phytoplankton in the water column to sustain the energy needs for egg production. The new generation develops in the surface water in close relation to the phytoplankton bloom. Growth in size takes place through successive moulting, first through the six naupliar stages and then through six copepodite stages. Stage CV is the normal overwintering stage. North of Lofoten, there is only one generation a year, whereas further south

Box 2: Phytoplankton model

The phytoplankton unit used in the model is: [mmol N m^{-3}], but the results are presented in chlorophyll units [mg Chl a m^{-3}] using a chlorophyll *a*:Carbon ratio of 6.3.

Transport and distribution of a scalar entity such as phytoplankton (P) or nitrogen (N) is governed by Eqs 2-4. Velocities and vertical mixing coefficients are taken from the hydrodynamical model. In the Barents Sea, ice is important for the light conditions in the water column. Ice cover is modelled by a ice model. The equations describing the growth of phytoplankton at a certain depth are:

$$\frac{dP}{dt} = PP^B_m \frac{Chl}{C}\left[\min\left\{1-\exp\left(\frac{-\alpha^B I_z}{P^B_m}, \frac{N}{k_N + N}\right)\right\}\right] - Pr - PS_{sed} \tag{10}$$

$$\frac{dN}{dt} = -PP^B_m \frac{Chl}{C}\left[\min\left\{1-\exp\left(\frac{-\alpha^B I_z}{P^B_m}, \frac{N}{k_N + N}\right)\right\}\right] + 0.5Pr \tag{11}$$

where P^B_m is the maximum photosynthetic rate [mg C (mg Chl)$^{-1}$ h^{-1}], *Chl/C* is the chlorophyll *a*: carbon ratio, α^B is the Chlorophyll *a* specific photosynthetic efficiency [mg C (mg Chl)$^{-1}$ h^{-1} (µmol m^{-2} s^{-1})$^{-1}$], k_N [mmol N m^{-3}] is the half saturation constant for uptake of nitrogen, r [h^{-1}] is the respiration rate, I_z is the irradiance at depth z [µmol m^{-2} s^{-1}], and S_{sed} [h^{-1}] is the sedimentation rate given by

$$S_{sed} = d_{mn} + (d_{mx} - d_{mn})\exp\left\{-\frac{N}{(k_N - N)/dg}\right\} \tag{12}$$

where d_{mn} [h^{-1}] is the sedimentation rate at high nitrogen concentration, d_{mx} [h^{-1}] is maximum sedimentation rate at low nutrient concentration and dg is a parameter which determines the functional relationship between sedimentation rate and concentration of nutrients.

The photosynthetic available irradiance (PAR) is calculated from the local height of the sun after a model by Bird (1984). Daily average cloud cover is interpolated from available meteorological stations in the Barents Sea. The depth variation of the irradiance is calculated after Kirk (1984):

$$I_z = \frac{I_0}{COSI}\exp\left\{\int_z^0 \frac{1}{COSI}\left(k_w + f_{att}(Chl)\right)d\tau\right\} \tag{13}$$

where I_0 is the irradiance at the surface, k_w is the attenuation coefficient of pure sea water [m^{-1}], *COSI* is the average cosine of light in water and $f_{att}(Chl)$ is a function that calculates the attenuation due to chlorophyll in the water column (Parsons & al. 1983):

$$f_{att} = 0.0088Chl + 0.054Chl^{2/3} \tag{14}$$

For more information, see Slagstad & Stokke (1994) and Sakshaug & al. (1995).

---

part of the population may have two generations. In July, the first overwintering animals are found below 600 m. During the winter, moulting into the last stage (CVI) female or male, takes place. *C. finmarchicus* overwinters in the Norwegian Sea at temperatures below 0 °C which usually are below 500 m. In the Barents Sea this depth is not found and we may find the overwintering stock at 200-300 m or even up to the surface (Pedersen & al. 1995).

C. glacialis is an Arctic species found in shelf areas which is seasonaly covered by ice. The life cycle is similar to *C. finmarchicus*, but the generation time is two years.

Zooplankton production

Attempts to model growth and secondary production have usually been performed as an investigation of the productivity of an area based on food availability (primary production), temperature and mortality due to predation. Such models have been used for the Barents Sea, but were not able to explain the variability in *Calanus* biomass that were observed in the 1980s (Slagstad & Støle-Hansen 1991). Model sensitivity analyses indicated, however, that the variability in the overwintering stock of *Calanus* would be an important factor determining the total production the following spring and summer (Slagstad 1981a, Slagstad & Støle-Hansen 1991). This has also been suggested in other studies (Colebrook 1985, Aksnes et al. 1989). Looking at a shelf ocean such as the Barents Sea, these would mean that there is a strong feedback from planktivorous fish to the secondary production. One of the major hypotheses drawn up after the PRO MARE was that import of *Calanus* from the Norwegian Sea could give an important contribution to the Barents Sea *Calanus* spawning stock.

Growth of a population (biomass formation) depends on the growth rate of each individual. Since weight specific growth and respiration rates of *Calanus* depend on the stage of development, the population structure (stage structure) should be a part of any model that intend to calculate population growth. The vertical distribution which is important for ocean-shelf transport and mortality rate, is also dependant on developmental stage. The *Calanus* model used here, therefore, consists of three parts: (1) a population model that calculates the stage distribution as a result of spawning time, developmental rate and advection (2) a physiological part that calculates hunger, growth rate and egg production rate as a function of temperature and concentration of phytoplankton and (3) a vertical migration model which computes

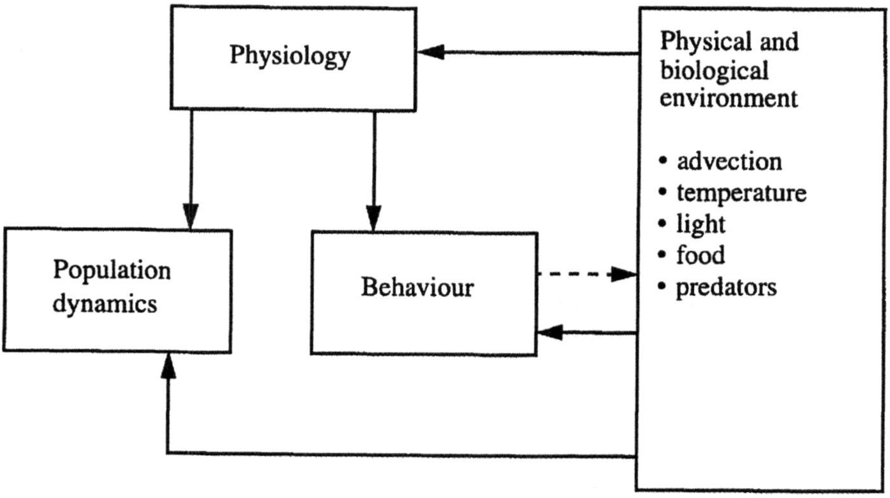

Figure 4. Interaction between physiology, vertical migration, population and environment in the *Calanus finmarchicus* model.

the vertical distribution as a dynamic function of light intensity and hunger. These three submodels interact with each other and the environment (Fig. 4).

Population sub-model

The population model consists of one part which produces a stage distribution of nauplii and one part which produces a weight distribution of the copepodites. An empirical relationship is used between stage and weight. The recruitment to the nauplii model is egg production from spawning of *Calanus* females. When the last nauplius stage is reached, the nauplii enters the copepodite model as CI. The copepodite model calculates the change in weight distribution as a function of growth rate, recruitment from the nauplii model and mortality according to the equation

$$\frac{\partial \eta(w)}{\partial t} + g(w)\frac{\partial \eta(w)}{\partial w} = R(w_0) - m(w) \tag{15}$$

where w is weight, $\eta(w)$ is number of individuals having weight w, g is growth rate for an individual of weight w, $R(w_0)$ is recruitment rate from the nauplii model and $m(w)$ is the weight dependent mortality rate. Growth rate is taken from the physiological sub-model. The mortality rate has a fixed value, depending on developmental the stage, only. This mean that longer time spent in a stage due to low temperature would increase the mortality. The equation is solved by a finite difference method, which for numerical reasons makes it necessary to divide each stage into several weight classes. A similar equation is used for the nauplii model.

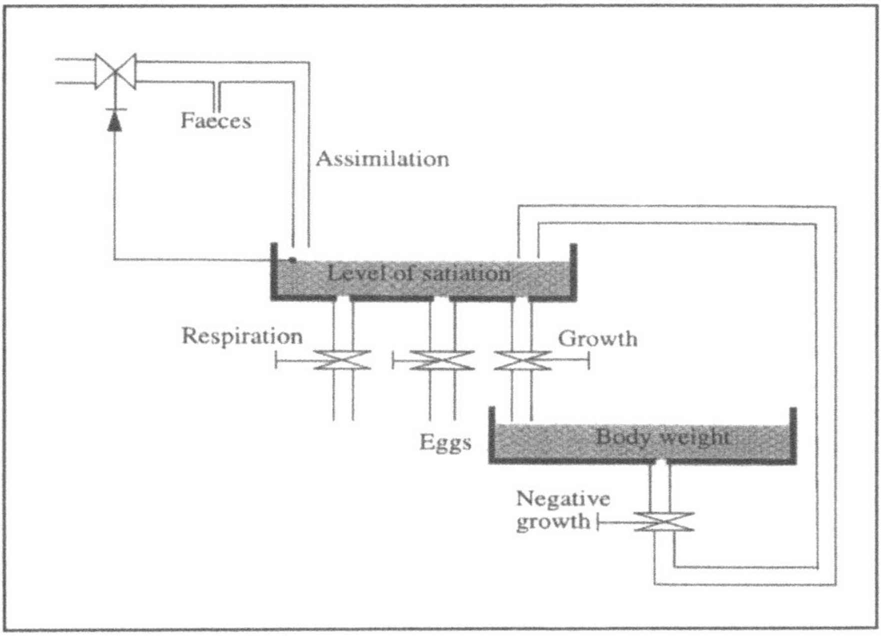

Figure 5. Illustration of the physiological model of *Calanus finmarchicus.*

Physiological sub-model

The physiological model has two state variables which are energy stores (Fig. 5). The first is simply energy stored in body structures and the fat droplet and is associated with energy which is not easily mobilised when energy is needed. The second state variable is associated with the energy store which is mobilised when energy is needed such as the pool of wax ester. The energy content of this store will determine the state of hunger and will affect the ability to migrate into the upper part of the water column even if the light intensity is higher than acceptable from a mortality point of view. This state variable (called *the level of satiation, s*) which is the inverse of hunger, will control both the filtering rate, growth and egg production rate. The state variables are governed by the equations

$$\frac{dw}{dt} = g = f(s, w, T) \tag{16}$$

$$\frac{ds}{dt} = \frac{[A(s,P) - g(s,w,T) - E(w,s) - R(w,T)]}{S_V} \tag{17}$$

where A is the rate of assimilation, g is the growth rate, E is the egg production rate,

R is the respiration rate, f is a function that gives the growth rate as a function of the level of satiation (s), weight (w) and temperature (T). P is the concentration of phytoplankton at the depth of the animal. The level of satiation is a relative measure for the energy content in this store as we divide by its maximum size S_V.

Vertical migration behaviour model based on physiological state

Copepodite stages of *Calanus* are able to choose its vertical position in the water column. This is important for the advection process from oceanic to shelf areas. Since the depth of overwintering in oceanic water is below 500 m, advection into shelf water must take place in the spring and summer. Since currents may have directions depending of depth, the vertical position in the water column is important for calculating the transport. Predation rate from visual predators may also be affected by the depth distribution (see the "Static optimisation" section and Box 5).

It may be argued that the largest copepodite stages should stay deep (and dark) enough to avoid being eaten by visual predators (see the "Age- and size-dependent variation in motivation" section). On the other hand, if this depth is below the phytoplankton layer, animals will not be able to sustain maximum growth rate and may not be able to reach the stage CV with sufficient energy reserves to make it through the winter. The optimal choice can be calculated by a procedure for dynamic optimisation (Fiksen & Giske 1995, "State dependency dynamic optimisation" section). One may, however, also argue that *Calanus* does not know anything about dynamic optimisation, but will through its sensory apparatus perceive the environment and combine this by its perception of its internal state (e.g. hunger) to decide what to do. The weighing of different sensory inputs (e.g. light intensity, predator level, concentration of phytoplankton etc.) will be a result of evolutionary processes which seek to maximise the reproductive output (cfr. the "Natural selection and maximisation of fitness" section). In the following, we will describe a vertical migration model which is based on physiological state and environmental variables.

The nauplii are assumed to be randomly distributed in the upper 40 m. The copepodites have an internal light reference level which will adjust their vertical position throughout the day. When the level of satiation is low, the light reference level is increased and animals move upward in the water column to search for food (i.e. they will tolerate higher mortality risk when hungry). After the level of satiation is filled up again, the light reference level is reduced and, depending on the ambient light intensity, the downward migration start. For a vertical migrating animal, the energy content in s will fluctuate, but the growth rate will only be affected when the level of satiation becomes too low

$$\frac{dz}{dt} = f_z(I_r, I_z) \tag{18}$$

$$\frac{dI_r}{dt} = f_{Iref^{(s)}} \tag{19}$$

where z is the depth in the water column, f_z is a function that calculates the vertical velocity from the difference in light reference level (I_r) and the ambient light at depth z (I_z), f_{Iref} is a function that calculates the rate of change in the light reference level as a function of s. In this version of the model, the concentration of visual predators is not taken into account, but it can be included in equation (19). More details about this model and parameter used can be found in Slagstad (1981b), Slagstad & Tande (1990) and Tande & Slagstad (1992).

Animal behaviour and life cycles

Proximate cues and ultimate forces

Both animals and plants react to environmental stimuli. That is why they have sense organs. Fishes have a wide variety of differentiated and specialised sense organs, making them capable of detecting changes in temperature, pressure, sounds, odours, light etc. In the study of the behavioural responses to the external environment, one can ask both *how* individuals sense and respond and *why* they do so. These are quite different questions: The *how* question will reveal which environmental factors triggers which behavioural response, i.e. the proximate cause of a response. The *why* question will reveal the ultimate cause of the sensitivity to the different environmental factors, i.e. why it is important to respond and which environmental factors one should expect the individual to have sensors to detect. To understand the ultimate *why*, one must focus on how evolution has formed the species.

Natural selection and maximisation of fitness

Natural selection is a genuine biological force operating with necessity when two conditions are fulfilled (e.g. Parkin 1979): 1) there is differential reproduction, i.e. not all individuals in a population produce the same number of offspring, and 2) the differences are heritable, or have a heritable component. After several generations in a habitat, natural selection will tend to make the individuals in a population similar in reproductive ability, manifested in local adaptations in body size, life history, reproductive capacity and behaviour. Evolution is therefore a good foundation for understanding characteristics of present living natural populations.

The force of natural selection has lately been demonstrated by the extensive use of genetic algorithms to solve complex numerical problems in science and engineering. The genetic algorithm kernel (GAK, Holland 1975, 1992) is a way of improving a "population" of suggestions for solutions to a complex problem by allowing small or large random changes in existing solutions ("mutations"), and to make new solutions by combining existing ("reproduction with recombination"). The performance of all suggested solutions are then evaluated, whereupon "natural selection" eliminates the poorest and reproduces the best, under a certain risk of mutations to allow for new ideas to come up. Thus, GAK will tend to improve solutions based on the existing alternatives, but will not necessarily find the globally best solution.

While natural selection will lead to the selection against inadequate responses, it is also necessary to show that behaviour to be selected for does exist in nature. This has been done in several laboratory investigations, where fish have been shown to be capable of performing optimisations in relevant situations (see Milinski (1986) for a review). The life history of a population, and not our suppositions, sets the limitations on what behaviour can be expected of an animal. Thus, Milinski (1979) showed that sticklebacks are able to assess feeding rates at different feeding locations, and respond optimal to experimental manipulations of feeding regime. Milinski (1985) showed that preferred feeding locations are influenced by physiological status and predation risk. Magurran & al. (1985) have shown that feeding or escape, i.e. the trade-off between predation avoidance versus feeding is influenced by shoal size. Utne & Aksnes (1994) have shown that the risk-willingness also depends on age — i.e. on the life-history stage of the individual.

The degree of adaptation to the environment, and the animal's ability to spread its genes to future generations, is the core of the fitness concept. Fitness can be defined as an individual's contribution to the next generation, relative to other individuals, and can be estimated by the difference in instantaneous reproductive rate of the individual (ρ) and the population (r):

$$\Phi = \rho - r \tag{20}$$

(Giske & al. 1993). To maximise Φ, an individual cannot do better than to maximise ρ, which therefore itself is a measure of optimal behaviour.

The development of biological optimality models started in 1966 by MacArthur & Pianka (1966) and Emlen (1966), who initiated the era of the "optimal foraging theory" (OFT, Schoener 1987). Laboratory investigations have generally supported OFT when animals were faced with simple set-ups (Stephens & Krebs 1986; Schoener 1987). OFT assumes that there exists a (linear) relationship between feeding rate and fitness (Calow & Townsend 1981). Implicitly, it was assumed that feeding could influence birth rate without also affecting death rate. Mangel & Clark (1986), however, showed that feeding alone cannot explain spatial distributions in general. By the use of dynamic optimisation, they showed that all fitness-related aspects of life could be combined in a common expression. At the same time, Life History Theory (LHT) developed from a descriptive to a predictive tool available to optimality studies (Stearns & Crandall 1984, Werner & Gilliam 1984, Sibly & Calow 1986).

Life history theory and static optimisations

Natural selection pushes a population towards better and better adaptations to the local environment, since the fittest individuals on average leave more and better offspring than the rest of the population in each generation. Such, natural selection acts to make all living beings into efficient "replicators", i.e. that individuals of any species who devote the most of their time and energy on leaving offspring will pass on this behavioural trait to coming generations. We can therefore assume that anatomy, physiology, life cycles and behaviour of organisms now occurring in their natural habitats are well adapted to this. And we can further assume that the characters of the individuals in an area are those that on average leave the most offspring. Hence, the "meaning of life" for any living creature is to maximise ρ in Eq. 20. But how should they then live?

The instantaneous reproductive rate ρ can be decomposed to three factors: 1) *fecundity*, the number of offspring produced by an individual, 2) *survival*, and 3) *time* (age). The reproductive rate can be expressed by the Euler-Lotka equation:

$$1 = \sum_{Y=1}^{\omega} m_Y l_Y \exp(-\rho Y) \tag{21a}$$

(Euler 1760, Lotka 1907) where m_Y is fecundity at age Y, l_Y survivorship from birth up to age Y and ω is last age of reproduction. Generally, fecundity and survivorship are interdependent. Early fecundity will be at expense of increased body mass and possibilities for even larger fecundity later in life. High food intake rate, which is needed to start reproduction at a minimum age, will often have a cost of larger exposure to predators. Natural selection will then work to optimise food intake, growth, fecundity and mortality exposure so that the resulting expected lifetime reproductive rate ρ is maximised. Repeated reproduction (several spawning periods) will develop when it does not pay to invest all reproductive capacity in the first attempt. For the case that an individual only reproduces once in a lifetime, the Euler-Lotka equation (21a) is simplified to

$$\rho = \frac{\ln(l \cdot m)}{T} \tag{21b}$$

where T is the generation time (age at reproduction).

If we can express how different environmental factors influence the life history variables l_Y and m_Y, then we may calculate the optimal behaviour of the individuals. From Eq. 21b we see that a favourable habitat is characterised by a potential for a high fecundity, a short generation time, and a low mortality rate. And although such a place is seldom found (and occupied if found), Eq. 21b shows how an optimal trade-off between these three factors should be performed to find the best of all available habitats. We see that factors influencing generation time has higher impact on fitness that factors influencing fecundity, due to the log-suppression of fecundity.

Werner & Gilliam (1984), Aksnes & Giske (1990) and Salvanes & al. (1994) have by different methods shown that an individual that may shorten its generation time by increased growth rate, should live as to maximise the ratio between somatic growth rate g and mortality risk M (maximise g/M). This is often the case for juvenile fish, while adults, which by definition already have reached the onset of reproduction, only can use extra food to increase their fecundity. For such individuals, the optimal trade-off between searching for food and hiding from predators shifts towards a more vigilant life. Giske & Aksnes (1992) showed that the optimum trade-off was to maximise $\ln g/M$, i.e. less emphasis on growth relative to mortality risk (see Fig. 6a). For many other aquatic animals, e.g. in major zooplankton groups, the major trade-off may be of mortality risk versus temperature (as high temperature is only found in the dangerous well-illuminated surface waters). This is illustrated in Fig. 6b, where the vertical distri-

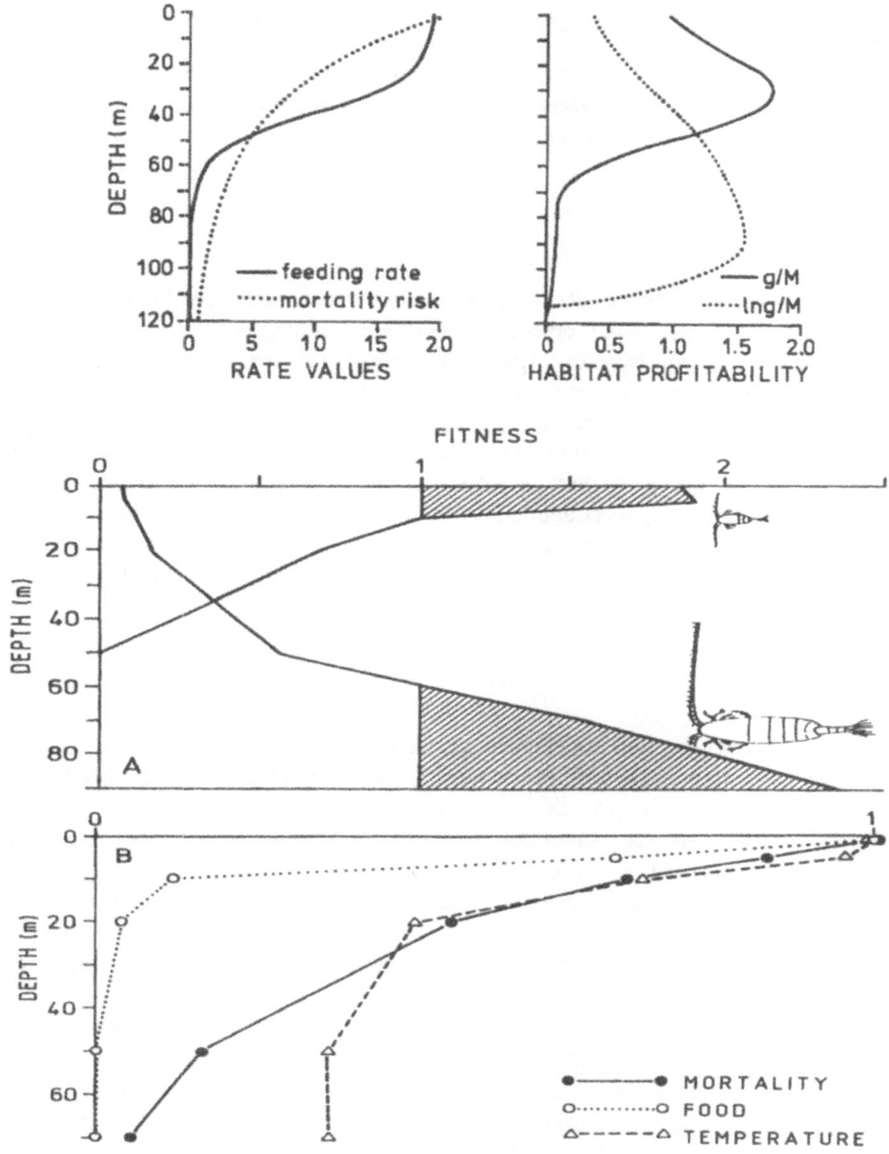

Figure 6. Top: Habitat profitability of fish. Juveniles may use extra growth to shorten time to maturation. Adults can only use extra food intake to increase fecundity. With typical vertical profiles of feeding rate and mortality risk (left), optimal depth will be shallower for juveniles (maximising *g/M*) than for adults (maximising ln*g/M*). Center: fitness value of depth differ for the small copepod *Paracalanus parvus* that may use warm water to shorten generation time than for the large *Calanus finmarchicus* that has only one generation per year. The environment is the same for both species and is shown below. From Giske & Aksnes (1991).

bution of two zooplankton species is explained by their different fitness value of growth versus survival, and by the impact of temperature upon generation time.

Dynamic optimisation

The attractivity of life history based models lies in their simple deduction from fitness and in their high degree of intuitive value. A problem is their coarse time scale; they will not resolve events on a short time scale. They may give average optimal solutions for juveniles and adults, and optimal solutions before and after spawning, but will not describe optimal trade-offs for hungry versus well fed animals or for adaptations to temporal and seasonal variability. Dynamic optimisation (or dynamic programming) is a method by which several time scales may be included, as well as the energetic status of the organism. This method is therefore far more computer-demanding than life history based models.

Dynamic optimisation has over the last decade been used in studies of optimal fisheries regulations and fisheries economy (e.g. Clark 1976, Charles 1983, Bjørndal 1988, Flåten 1988, Bjørndal & al. 1993, Hanneson 1993). State variables are used to examine the consequences of an action in fitness-related terms in order to model optimal behaviour (Houston & al. 1988; Mangel & Clark 1988). A decision variable (e.g. optimal habitat) is made dependent on the internal state variables (e.g. physiological state) and external time-dependent variables (e.g. food availability, predation risk and spawning season). Dynamic optimisation is a three step process. First, optimal decisions for all states at all times are found by backwards iteration from a fitness-related function of the state variable at the end of the modelled period. Secondly, the ultimate state distribution of the population is found from an initial distribution, assuming optimal behaviour of all individuals. Finally, the fraction of the population adopting a particular behaviour is found by combining steps 1 and 2. (See Box 3 for technical details.)

The dynamic optimisation model will generate mortality rates for animals adopting the optimal strategy, and feeding and growth rates of the population can also be calculated in the last part of the three-step process. Since fisheries contribute heavily to mortality of capelin in the Barents Sea, geography, intensity and timing of fisheries could be included in the mortality risk (c.fr. Law & Grey 1989; Sutherland 1990; Jensen 1991). This has not been done so far.

Box 3: Dynamic Programming

Finding optimal distributions by backward iteration proceeds as follows: relative fitness is first determined for each possible energetic state at the end of the simulated period, e.g. by a life history related function. Generally, these fitness values are termed $F(x,t,T)$, where x is the value of the state variable, t is current time step number and T is total number of time steps.

During a time step, an animal may choose among several habitats. Predation risk (p), probability of finding food (e), the energetic value of the food (f) and metabolic costs (c) may differ among habitats. (For simplicity of presentation, we here assume that all prey in a habitat are equal and that one or no item can be caught in a period.) In a period (from t to t+1) an animal of state x (x,t) either finds food and changes its state to $(x\text{-}c\text{+}f,t\text{+}1)$, or does not find food, and then changes to $(x\text{-}c,t\text{+}1)$. The average new state (x'_{t+1}) for an animal of state x_t in a habitat is therefore

$$x'_{t+1} = e(xt\text{-}c\text{+}f) + (1\text{-}e)(xt\text{-}c) \qquad (22)$$

and the probability of surviving the time step is (1-p). The fitness value of the states x-c+f and x-c at final time (t = T) is now already found, and fitness of state x in habitat h at t = T-1 is

$$F(x,h,t{=}T\text{-}1) = (1\text{-}p)\,[eF(x\text{-}c\text{+}f,T) + (1\text{-}e)F(x\text{-}c,T)] \qquad (23a)$$

$F(x,h,t{=}T\text{-}1)$ is then calculated for all habitats, and the optimal habitat is where $F(x,h,t)$ is maximal. In this optimal location $[h(x,T\text{-}1)]$ the fitness value of animals of state $(x,T\text{-}1)$ is

$$\mathrm{F(x,T\text{-}1) = (1\text{-}p^*)\,[e^*F(x\text{-}c^*\text{+}f^*,T) + (1\text{-}e^*)F(x\text{-}c^*,T)]} \qquad (23b)$$

where * refers to environmental values in the optimal habitat. Having found $F(x,T\text{-}1)$ and $h(x,T\text{-}1)$ for all x, calculation of fitness values and the corresponding optimal locations for all states at t = T-2 proceeds according to the general equation

$$F(x,t) = (1\text{-}p^*)\,[e^*F(x\text{-}c^*\text{+}f^*,t\text{+}1) + (1\text{-}e^*)F(x\text{-}c^*,t\text{+}1)] \qquad (23c)$$

Density dependency: Ideal Free Distribution

The models referred to above describe optimal trade-off for the individual, since the individual is the unit of selection. The value of a habitat for an individual does however depend on the presence of other individuals, as demonstrated by Magurran & al. (1985). Group size will influence both feeding (Clark & Mangel 1986) and mortality (Milinski 1986; Jakobsen & Johnsen 1988). Schooling and shoaling may increase individual feeding opportunities as the search volume of the group is larger than of the individual. Where food is patchy distributed, and especially if patches are ephemeral, the benefits of group searching will outweigh the costs of resource sharing (Clark & Mangel 1986). Under other circumstances, e.g. with overlapping search fields at low prey densities or for bad positioned individuals, flocking may reduce individual feeding rates. The fact that shoaling and schooling still are maintained, suggests that flocking has other functions than just to increase feeding. If predators feed by eating single food items, prey patching may reduce prey mortality risk both by dilution (Milinski 1977) and confusion (Milinski 1984). These effects also allow schools of fish to feed in areas where mortality risk for single individuals would be too high. Each individual in a school may also spend less time and attention on predator avoidance, and schools will generally discover a predator earlier than solitary individuals. Due to confusion costs, planktivores feeding in dense zooplankton swarms have a lower ability to locate their own predators (Milinski 1986). Unless they are very hungry, planktivorous fish therefore tend to feed at the edge of swarms, where feeding rate, confusion costs, and mortality risk is lower (Milinski 1977; Godin & Smith 1988).

So, while dynamic optimisation increases realism in behavioural models by including internal state and variable motivation, the method treats interactions between individuals in a static manner. A different concept is the density-dependent life-history based model, where habitat quality is directly linked to competitor density. Group-size dependent optimal distributions within and among feeding sites have been studied using the Ideal Free Distribution (IFD) concept (Fretwell & Lucas 1970; Fretwell 1972). It is called 'free' because all individuals are assumed free to relocate at no cost, and it is 'ideal' because each individual is supposed to move to the site where its gain will be highest. The original IFD was based on four assumptions: 1) a number of equal competitors are found in a habitat, 2) within the habitat there are several different patches of food available to the competitors, 3) the competitors can move freely at no cost among these patches, and 4) each individual will go where its expected gain is highest. The competition was further assumed to be 'scramble', without any contest or combat between the competitors. The theory of Ideal Free Distributions (IFD) has been used by MacCall (1990) to distribute fish optimally among unequal patches. IFD was developed in the era of optimal foraging theory, and although Fretwell & Lucas (1970) stated that food was but one component of fitness, IFD assumed a direct relationship between feeding rate ('gain') and fitness. We have been interested in IFD as an alternative to dynamic programming, as the latter focuse on internal state as basis for motivation, and then cannot easily include effects of competitors. [Some dynamic programming models investigate optimal groups size, but then at the expense of physiological state (Mangel 1990)].

For our purpose, the IFD had to include density-dependent effects of predation risk, also. To incorporate also mortality risk, the second assumption may be restated: 2) 'Within the habitat there are available to the competitors a continuous range of localities differing with respect to expected fitness gain'. (See Box 4 for technical details.)

The assumption of 'free' travel between patches may hold for small distances. However, Tyler & Gilliam (1995) showed that swimming costs had to be considered for the distribution of stream fish. Even more will such costs apply to long-distance horizontal migrators. Also the assumption of 'ideal' individuals with perfect knowledge is questionable, both for use in IFD models and elsewhere. Several authors have found that perceptual constraints may reduce the ability of the fishes to assess the quality of their habitat so that the spatial distribution was more random and less ideal (Abrahams 1986, Utne & al. 1993, Gray & Kennedy 1994, Tyler & Clapp 1995).

Box 4: Ideal Free Distribution

Under Ideal Free Distribution with equal competitors, the total number of individuals distribute among the habitats so that the gain of an individual is equal in all habitats. In the classical version of the model, food was the resource, and the individual gain

(S_i) in a habitat (i) declined from a density-independent maximum B_i by a habitat-specific function of competitor density d_i

$$S_i = B_i - f_i(d_i), i = 1, 2, ..., L \text{ (habitats)} \tag{24}$$

and the IFD is achieved when individual gain is equal among habitats

$$S_1 = S_2 = ... = S_i \tag{25a}$$

and total number in the $K \leq L$ occupied habitat match population size N_T

$$N_1 + N_2 +++ N_K = N_T \tag{25b}$$

For zooplankton and fish in a natural environment it is not to be expected that food gradients alone shall cause an IFD, for three reasons: 1) there are combinations of low competitor density and high food production where individual feeding rate will not be influenced by a (minor) increase in competitor density, and more important, 2) one of the main resources for development and fitness for aquatic organisms may be environmental temperature and 3) predation risk will affect fitness in a density-dependent manner. Habitat profitability may then be expressed directly in terms of the habitats' expected contribution to the fitness (Φ) of the animal (Eq. 20), and ideal free individuals will distribute so that

$$\Phi_1 = \Phi_2 = ... = \Phi_i \leftrightarrow \rho_1 = \rho_2 = ... = \rho_i \tag{26a}$$

[as r is a population parameter constant for all habitats (and is the average of all ρ_i's)]. In a situation where feeding will be sufficient for growth and where reproduction may occur unhindered by e.g. seasonal constraints, fitness is proportional to the life-history trade-off g/M, as shown above. Then we may write

$$\Phi_1 = \Phi_2 = ... = \Phi_i \leftrightarrow g_1/M_1 = g_2/M_2 = ... = g_i/M_i \tag{26b}$$

(Giske *et al.* 1997). In IFD terms we will find the relation between the optimum number in each habitat, so that gain is equalised by competition and predation risk dilution. Fecundity, growth and survival will be derived from mechanistical equations of the environment and density-dependent responses. The ideal free distribution — where no individual could increase its fitness by moving — is obtained when competitor density in each available habitat is adjusted so that the ratio g/M is equalised.

ECOLOGICAL MODELLING OF THE BARENTS SEA

Physical oceanography

The first attempt to develop a 3-dimensional (3D) hydrodynamical model for the Barents Sea took place in 1975 when the OCEAN-BIO_MODELS research program started (Slagstad et al. 1975, Balchen 1980, Berntsen et al. 1981). The model was not completed with open boundaries. Parallel to this work, a 1-D (vertical direction) was used to simulate primary and secondary production in different water masses (Slagstad

& Tande 1990, Slagstad & Støle-Hansen 1991). This model used measured sea surface values of temperatures and ice covers combined with the general seasonal knowledge of seasonal variation to calculate vertical mixing coefficients.

In 1987 the first 3D-model using the main driving forces was applied for the Barents Sea (Slagstad 1987). The model was used to calculate the hydrodynamical environment (flow field, vertical mixing coefficients and ice cover) to be used in a 3D phytoplankton model (Slagstad & Støle-Hansen 1991, Støle-Hansen & Slagstad 1991). This work was a part of PRO MARE and has also continued during the Multispecies program (Slagstad & Stokke 1994).

Other models have been used to study variations in wind driven inflow of Atlantic water into the Barents Sea (Ådlandsvik & Loeng 1991) and drift of fish larvae from the spawning grounds along the Norwegian Coast to the Barents Sea. At The Norwegian Meteorological institute, a 3D model is running operationally. Study of heavy deep water formation in the Barents Sea has been performed by a so-called plume model by Harms (1994).

The circulation pattern in a shelf ocean like the Barents Sea are governed by the prevailing wind, flux through the boundaries and the density field. Ådlandsvik & Loeng (1991) showed that the variations in the inflow through the Fugløy-Bear Island section was mainly due to the wind. Easterly wind tends to slow down the transport from the Norwegian Sea whereas westerly wind increases the inflow.

Vertical mixing is important for the primary production. In the spring, the deep winter mixing brings nutrients into the surface water, but the average light intensity in the water column is too low to allow any net phytoplankton growth. In oceanic waters away from the coast or other sources of fresh water influx, the water column has to be stabilised by thermal heating of the surface layer by solar input or heat transfer from the atmosphere. Wind tends to break down the stabilisation, especially wind from north which usually is cold and thereby decreases the vertical stability by vertical convection in addition to the mechanical steering. If fresh water is added to the sea surface, stabilisation may take place independent of the heat flux from the atmosphere. Large areas of the Barents Sea are dominated by fresh water input from various sources. Along the ice border melt water is formed when the ice comes in contact with warm Atlantic water. Since most of the salinity contained in the water is rejected during the formation of the ice, fresh water is released when the ice melts. This creates a thin stable layer of low salinity water along the ice border. Along the coast, the Norwegian Coastal Current and freshwater run-off from land create a layer of low saline water which stabilises the water column.

Model set up

The hydrodynamical model (Box 1) is implemented for an area from SW Norway to the Kara Sea (Fig. 7). Horizontal grid point distance is 20 km. The vertical column is divided into 20 levels (layers) of different thicknesses; from surface and down: 10 m, 6 x 5 m; 10 m; 6 x 25 m, 2 x 50 m; 100 m, 200 m, 400 m and 1000 m. The initial density field for an autumn situation was taken from the Climatological Atlas of the World Ocean produced by the National Oceanographic Data Centre (NODC) in USA (Levitus 1982). In the Barents Sea this field has been improved by measurements performed by the

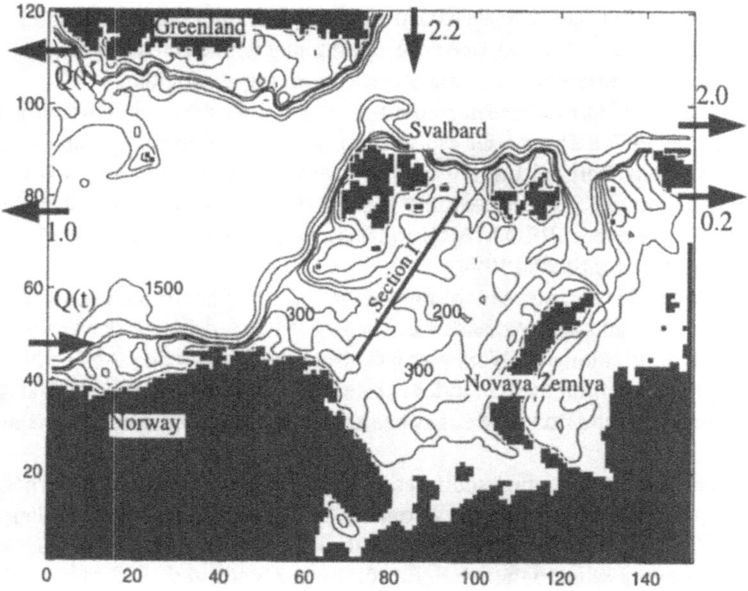

Figure 7. Model area and specified flux through the open boundaries in Sverdrups ($10^6 m^3 s^{-1}$). Isobaths are shown for 100, 200, 300, 400, 500, 1000, and 1500 m. "Section I" indicates a transect from where data in Figs 10 and 12 are presented.

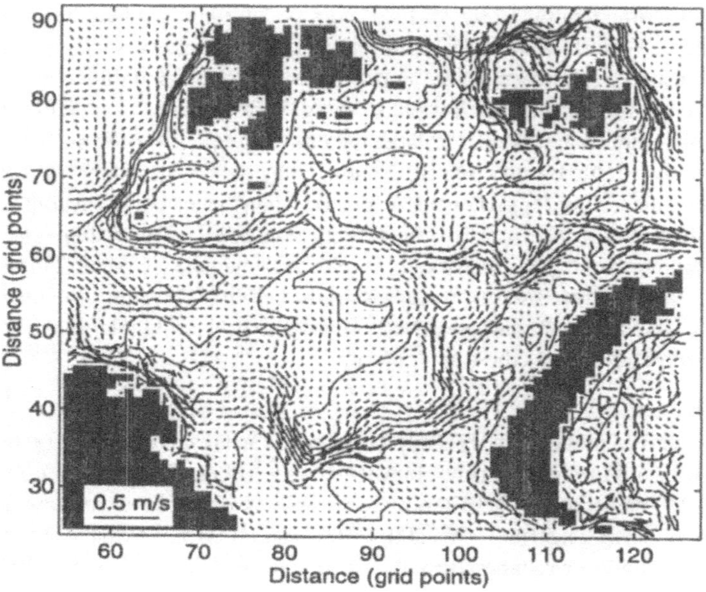

Figure 8. Simulated average surface currents for June 1984.

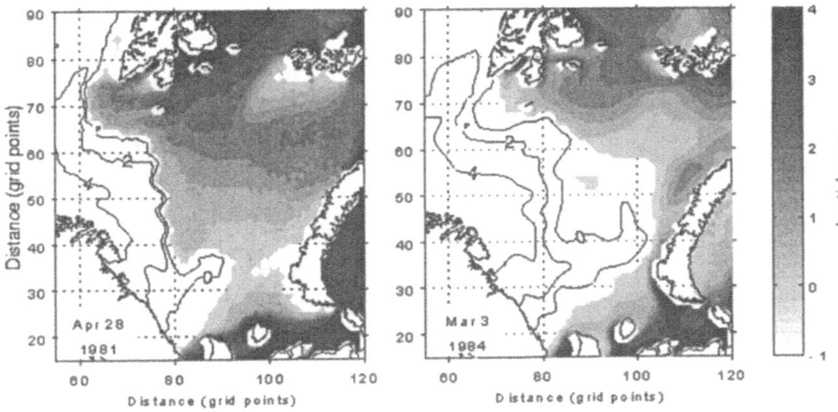

Figure 9. Simulated maximum ice cover and surface temperature for the "cold" year 1981 and the "warm" year 1984.

Institute of Marine Research in Bergen. The flux along the Norwegian Coast (Q(t)) is calculated according to an algorithm by McClimans (1993). More information about the model and the model parameters used can be found in Slagstad (1987) and Slagstad & Stokke (1994).

Results

When the effect of wind and tidal forces are smoothed out by averaging the flow field over a period of one month, we observe that the water movements tend to follow the depth isolines (isobaths) (Fig. 8). This mechanism, often referred to as topographic steering, often dominates the average flow pattern on shelf areas with strong variations in the bottom topography.

Extensive hydrographical data collection are usually performed in the Barents Sea each autumn. In winter and spring the available hydrographical data are more scattered especially in the ice covered areas. In order to create an environment which can be used in connection with the biological model runs, simulations have been initiated with a measured distribution of temperature and salinity. This field was advected from autumn to the spring using meteorological forcing (wind, air pressure, air temperature, humidity and cloud cover). The daily average flow field was stored and used for advection scenarios shown in this paper. Simulated maximum ice concentration in late winter is shown in Fig. 9 both for a cold year with heavy ice cover in the southeastern Barents Sea and for a warm year with much less ice cover.

The simulated hydrographical sections show the same vertical structure as the data (Fig. 10). Along the ice border, the melting ice creates a zone of high vertical stability. This stability is also found during the summer behind the ice border. South of the ice border, the water column in well mixed in April, but the depth of the vertical mixing is gradually reduced due to reduction in thermal heating and wind speed.

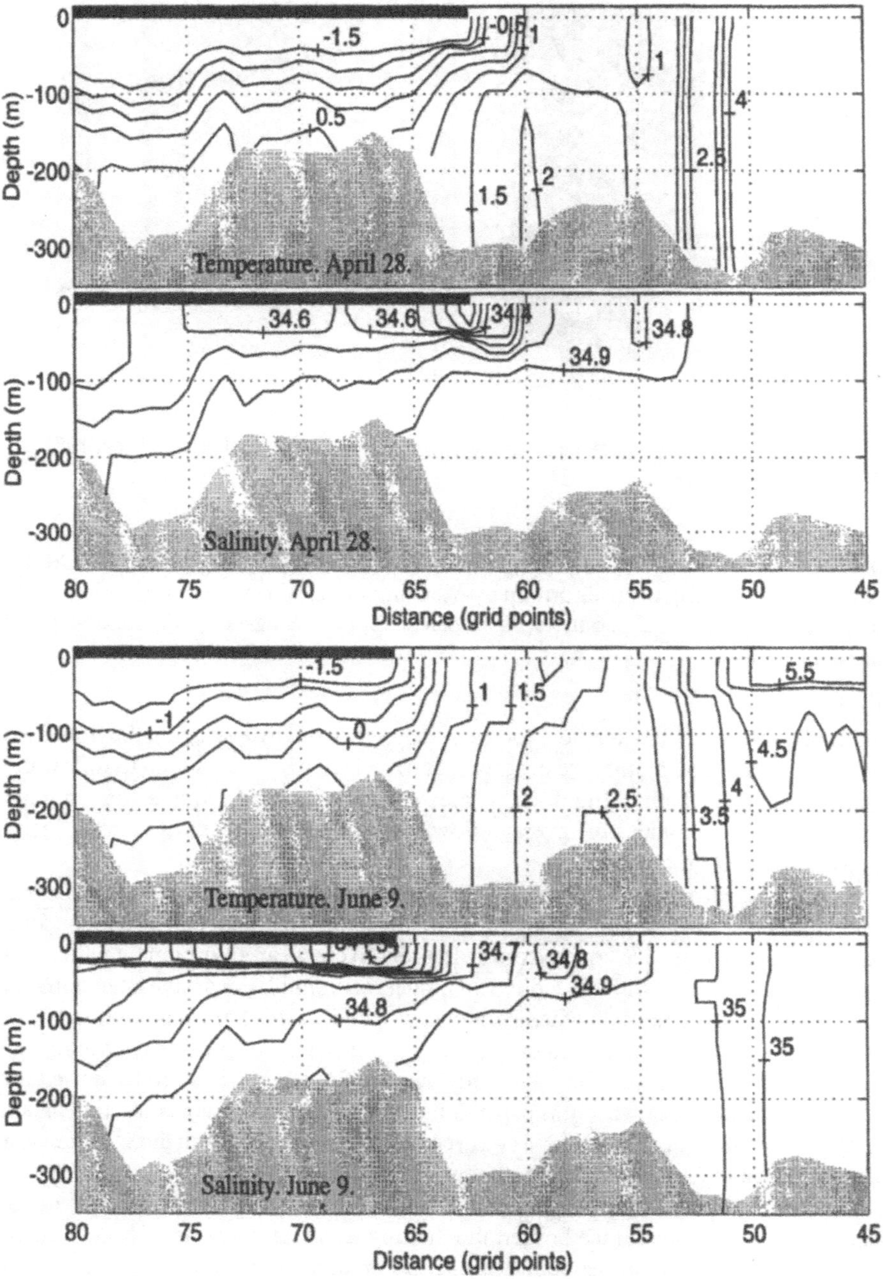

Figure 10. Simulated vertical distribution of temperature and salinity along "Section I" in 1984 after a simulation period of 210 days (upper) and 252 days (lower). The position of the section is shown i Fig. 7.

Primary production

The simulation start for phytoplankton is March 1st, driven by the data from the hydrodynamical model. These data were: daily average of three velocity components, vertical mixing, temperature, ice thickness and ice concentration. The initial concentration of phytoplankton and nitrogen (nitrate) were 0.2 mg Chl m^{-3} and 11 mmol N m^{-3}, respectively, in the whole model domain.

Simulated surface concentration of chlorophyll for the warm year 1984 is shown in Fig. 11. In the beginning of May the concentration of chlorophyll is rapidly increasing, especially along the ice border. The south eastern Barents Sea has the earliest bloom due to meltwater input which stabilise the water column. In late May most of the ice-free areas now has a bloom, except for the Bear Island Trench, where the bloom does not start until June. This pattern is similar from year to year. After the nutrients in the surface are depleted, further primary production will take place deeper in the water column. In the northern areas a surface bloom will take place as soon as the ice concentration is low enough to permit light to penetrate into the water column.

Effect of ice on primary production

To illustrate the difference in primary production between a warm and a cold year, the primary production for the cold year 1981, with extensive ice cover during the late winter has been simulated. The annual primary production along "Section I" was much higher in 1984 than in 1981 (Fig. 7). The greatest difference was found in the Atlantic water, west of the Central Bank. The 1981 production here is less than half the production of 1984 (Fig. 12). The simulated total primary production for the Barents Sea was about 30% higher in the warm year 1984 than in year 1981 (Fig. 13). Most of this difference is due to the extent of the ice cover. Areas covered by ice in the late winter will have a very stable water column the following spring and summer, inhibiting nutrients to enter the euphotic zone by vertical eddy diffusion.

Zooplankton dynamics

The ultimate goal of a coupled *Calanus* model is to simulate horizontal (and perhaps vertical) distribution and stage distribution which can be compared with measurements. Knowledge or estimates of future physical forcing and predation from planktivorous fish should make us able to predict future development in the *Calanus* biomass and distribution. However, more spatial and temporary resolved data are needed especially for obtaining an initial distribution of the overwintering stock. However, a model without such input data can be a useful tool to investigate the factors that affect the distribution and biomass formation of *Calanus*. Here we shall examine possible consequences of the import of *Calanus* from the Norwegian Sea.

Inflow of C. finmarchicus *from the Norwegian Sea*

To investigate the effect of the inflow from the Norwegian Sea we have made two model runs, one with an initial stock of females only on the Barents Sea shelf and one with an even initial distribution of females in the whole model domain.

The ontogenetic migration behaviour is such that females ascend to about 250 m in the first half of March and stay there until the concentration of phytoplankton is high

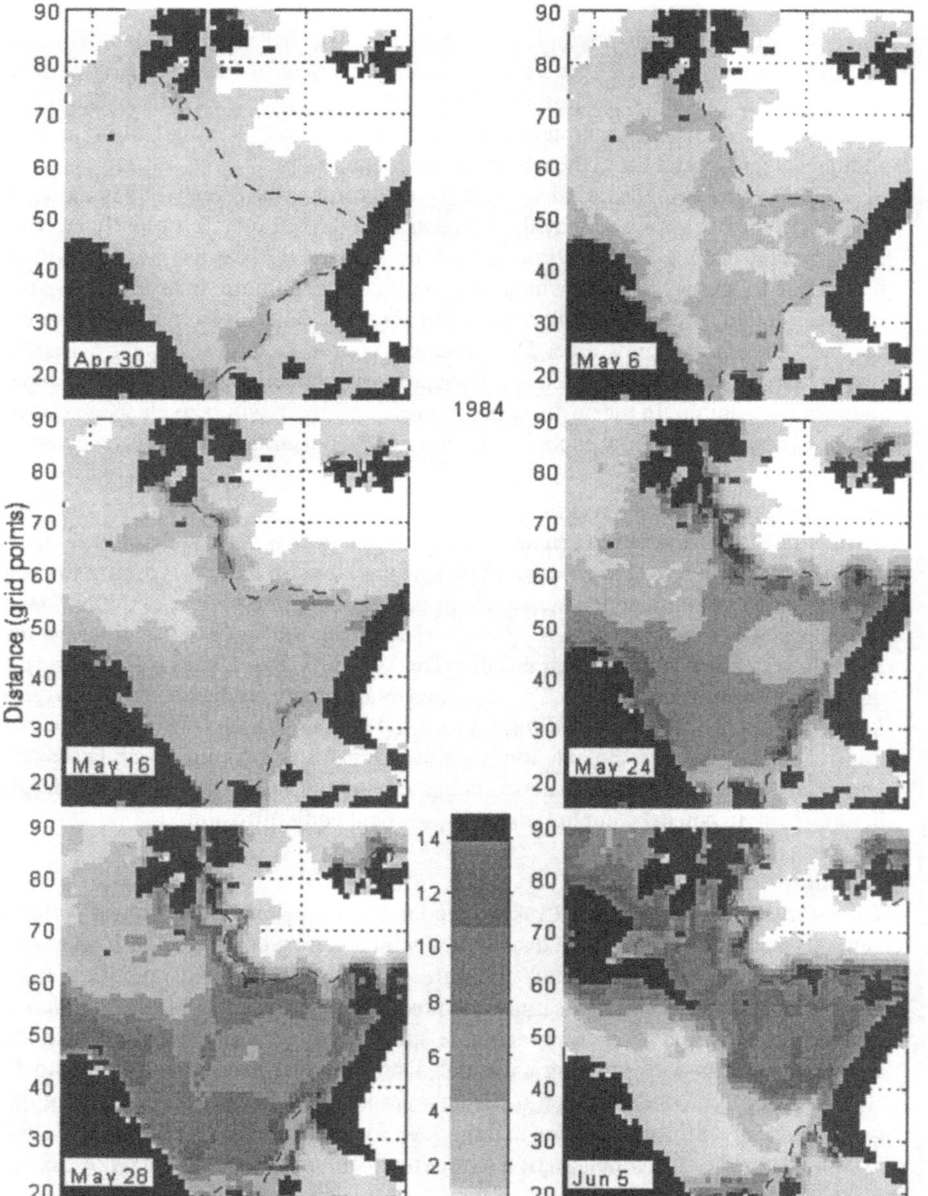

Figure 11. Simulated concentration of chlorophyll (mg Chl m^{-3}) at the surface at selected dates spring and summer 1984. (The figure continues on the next page).

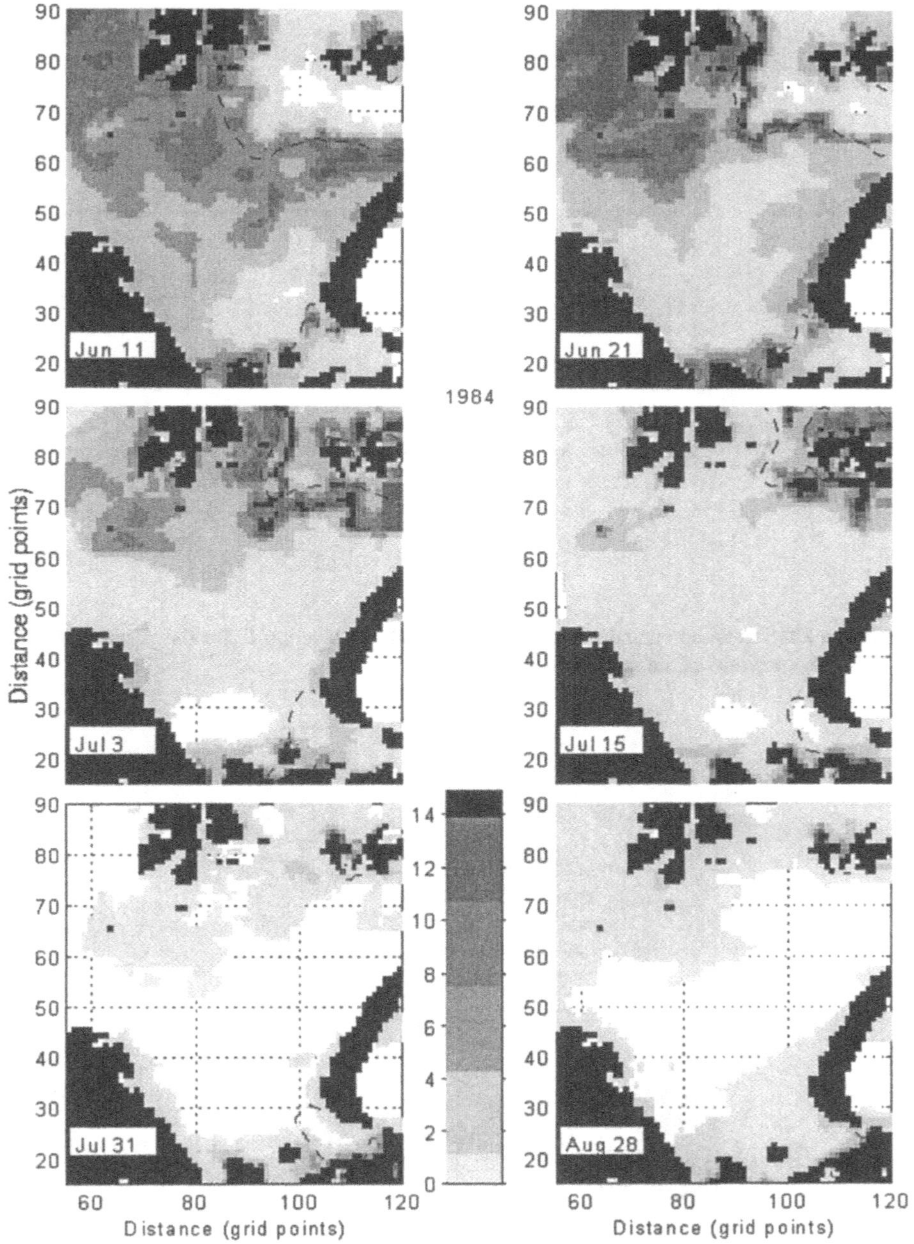

Figure 11 continued from previous page.

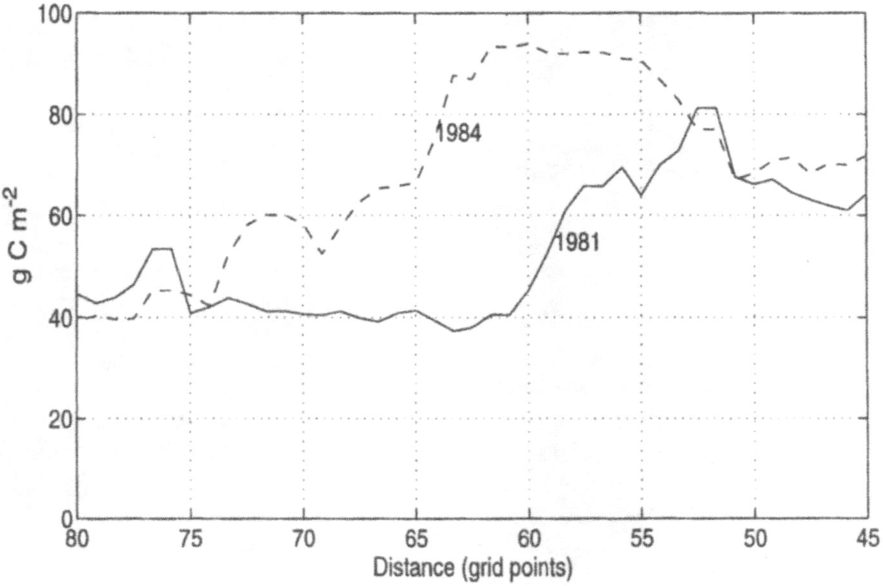

Figure 12. Simulated annual primary production along "Section I" for the years 1981 and 1984. The position of the section is shown i Fig. 7.

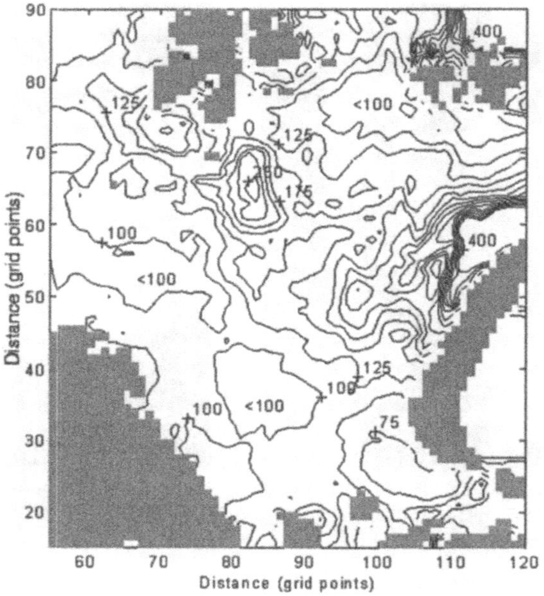

Figure 13. Relative annual primary production in 1984 in percent of the production in 1981.

enough to sustain egg production after which the females enter the surface layer. The vertical migration behaviour of the copepodites are calculated according to Eqs 18-19. When the last weight class within the Stage CV is reached, downward migration to 500 m is assumed. If such depth is not found, the CV's migrate towards the bottom.

In the first model run, we assume no *Calanus* in the inflowing water. The average overwintering stock is 2000 females m^{-2}. The physical forcing is from the relatively warm year of 1984. The new generation develops fast in the South and a shifts gradually northwards as a result of the decreasing temperature to the North (Fig. 14). Stage CV are first seen at the end of June East of Vardø. At the end of August, the CV are found in the South East basin and in the deep areas South East of Nordkappbanken. Very few stage CV are found north of the Polar Front. The average concentration of Stage CV on the Barents Sea shelf in late August was 700 ind. m^{-2}.

In the second run we assume 2000 females m^{-2} in the whole model domain (Fig. 15). The first CI are seen on Tromsøflaket at the end of April. These animals are born in the Lofoten area. In late May CI is found in the whole South Western Barents Sea (Fig. 15a). In June the first overwintering stages tend to concentrate in deep areas, west of Tromsøflaket and Ingøydjupet. This is emphasised later in the summer (Fig. 15c). The average concentration of CV in late August was 7960 ind. m^{-2}.

A third simulation staring with initial distribution of females only in the oceanic Norwegian Sea produced an average concentration of 2760 ind. m^{-2}. Thus, import from the oceanic regions of the Norwegian Sea alone produces 4 times the biomass of CV in the Barents Sea in August compared with the nonimport case. When the initial distribution of females were placed in the whole model domain, the biomass of CV was 11 times the nonimport case. The reason for this high number is the production on the Mid Norwegian shelf. Copepodite stages from this area are transported north with the Norwegian Coastal Current. The animals that have reached the overwintering stage will tend to be trapped west of Tromsøflaket when they descend to depths below 500 m, but those who are still in their growth phase and most of the time near the surface, will be advected into the Barents Sea.

The processes that dominates the amount of *C. finmarchicus* found in the Barents Sea seem to be:

1. The strength of the physical transport from Norwegian Sea (Lofoten Basin).
2. Size of the overwintering stock along the North Norwegian shelf break.
3. Spawning and survival success south of Lofoten (on the Mid Norwegian shelf and along the shelf break).

Models of motivation and behaviour

Static optimisation

In the following three examples on the applicability of static optimisation will be provided. The central assumption for all models is that the general Life History Theory can account for the optimal trade-off between opposing forces: growth versus survival or current versus future reproduction.

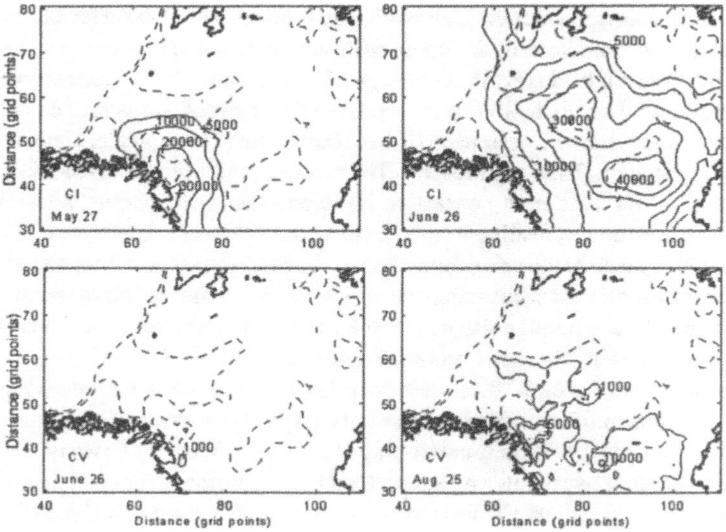

Figure 14. Simulated distribution of *Calanus finmarchicus* when the initial concentration of females are found in Barents Sea, only. A and B: Stage I (1000 individuals m^{-2}) on May 26th. and June 26th. and C and D: Stage V on June 11th. and August 25th.

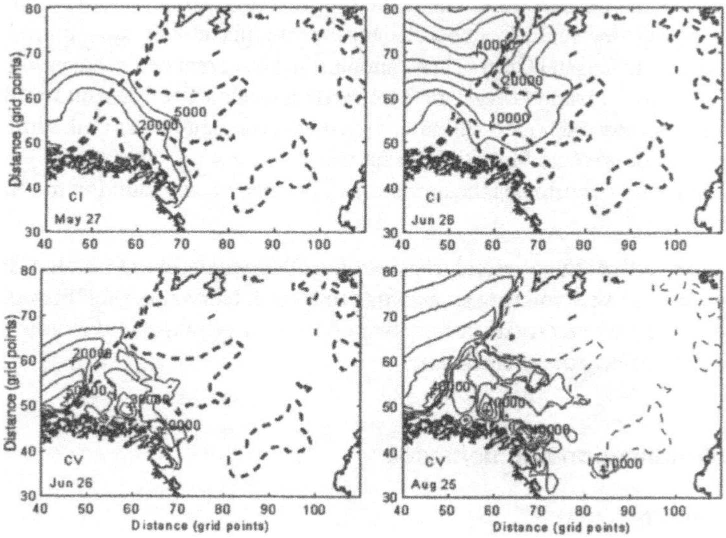

Figure 15. Simulated distribution of *Calanus finmarchicus* when the initial concentration of females are found in the whole model domain. A and B: Simulated distribution of *C. finmarchicus* stage I (1000 individuals m^{-2}) on May 26th. and June 26th. and C and D: Stage V on June 11th. and August 25th.

Life cycles of herring and capelin

The two major planktivores in the Barents Sea are capelin and juvenile herring. Capelin spends its entire life in the Barents Sea, while herring eggs are spawned at the west coast of southern Norway, the larvae drift into the Barents Sea, and adults live in the Norwegian Sea. The life cycles of these species are quite different. Capelin reproduces once, while herring may survive 15-20 reproductive seasons. Why is it so?

One can argue (Eq. 21b) that the optimal habitat for any organism is where the generation time is at a minimum, which at these highly seasonal latitudes is one year for fish. If this is achieved, the next goal is to maximise survival probability until the spawning season. In this life history strategy, fecundity is not of great importance, and egg numbers will count hundreds rather than millions. Due to the vertical correlation between feeding opportunity and mortality risk through the exponential decay of light (see explanation in Box 5), they will often be found deep in the water most of the year and the individual production of such animals will be low. Many fishes in the mesopelagic assemblage fit into this category. However, the Barents Sea is not a good environment for mesopelagic fish, as the bottom depth is too shallow to keep survival rate high, and because the long period with continuous daylight also enables predators to maintain continuous high activity.

Another life cycle with a higher age at maturity, lower juvenile survival and higher fecundity is common among long-distance migrants. Long-distance migration is very costly for small-bodied forms, and this life style thus requires a rapid juvenile growth. A large body size is also needed to have a potential for a high fecundity. Such animals will generally have high individual growth rate and potentially a high population biomass. Fishes in this life style are valuable resources to man, and include all the large stocks exploited in the Nordic Seas. This distinction between the mesopelagic, short-lived, low-fecundity life form and the horizontally migrating, high fecundity life form with multiple spawning seasons corresponds to the life history types Ia and Ib of Kawasaki (1980, 1983). Both types are typical in boreal waters. Kawasaki (1980, 1983) argues that both types are evolutionary adaptations to variable and unpredictable environments. Type Ia is adapted to irregular variation in environment and fecundity by a potential high per capita reproductive rate ρ, while type Ib is an adaptation to long-period variation, by high longevity and high adult survival. Capelin does not fit into any of these simple categories, as it is a long-distance migrator with only a small fraction of the population having more than one spawning season.

Life histories with drifting planktonic larvae which are susceptible to fluctuations in currents and feeding conditions, are likely to experience large fluctuation in juvenile survival. Most of this mortality takes place in the first week after hatching, in relation to start-feeding (Miller & al. 1988, Bailey & Houde 1989). This is related to the reproductive strategy of producing many small offspring. To counter the effect of variable juvenile survival the fish may spread the risk over several spawning seasons (iteroparity) and spread the eggs over a prolonged period (batch spawning) (Armstrong & Shelton 1990). Murphy (1968) and Kawasaki (1980, 1983) concluded that high fecundity, iteroparity and delayed maturation would increase juvenile survival in environments with variable juvenile mortality. Nevertheless, there are typically large

Box 5: Light as forcing for feeding and predation risk

Light does not only influence the ecosystem through primary production, but makes also a direct and instantaneous effect at higher trophic levels. Traditionally, prey concentrations have been regarded as the central variable for feeding opportunities. Visual predation is an important feeding mode for pelagic fish, and the highly variable surface light, the exponential attenuation in water, and its dependence on the variable scattering and absorbance properties of the water, makes the feeding process and thereby the mortality risk highly dependent on the light regime. Light intensity will thus influence the growth, mortality and spatial distribution of zooplankton, planktivores and fish-eaters in the Barents Sea. As many of the examples given in this section are vertical models, we will illustrate how light influences prey detection (Aksnes & Giske 1993).

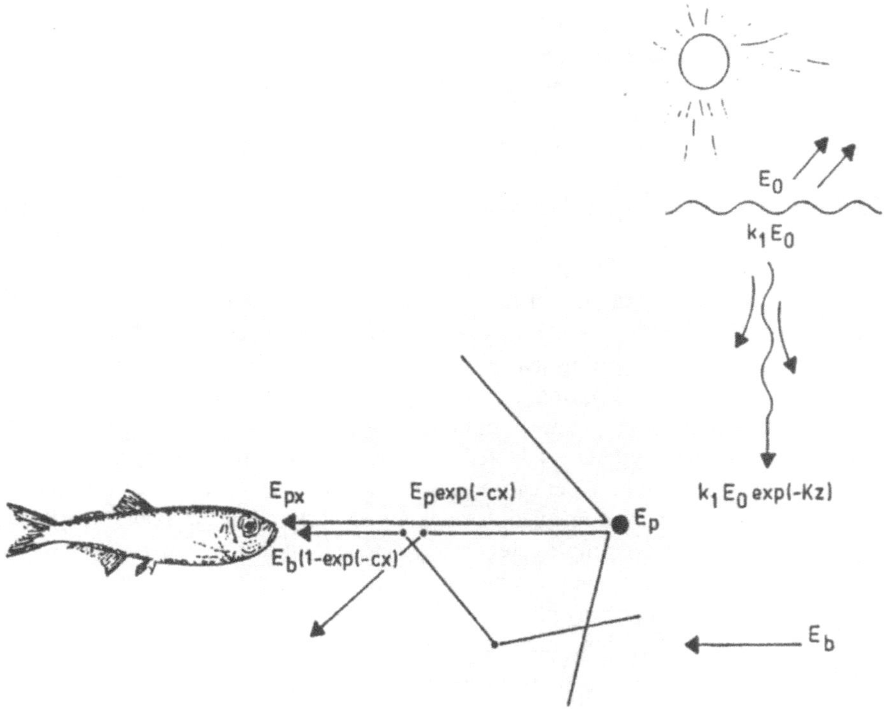

Figure 16. Simplified aspects of fish vision (after Skjoldal & al. 1993). Surface light in the air (E_0) is reflected and absorbed at the air-sea interface (ρ). Downwelling light is reflected and scattered (K) so that light at depth z [$E_b = \rho E_0 exp(-Kz)$] is diffuse. Irradiance reaching the fish eye has two sources: image-forming light from the prey [$E_p exp(-cx)$] where c is the beam attenuation coefficient and x is the horizontal distance, and scattered light [$E_b(1\text{-}exp(-cx))$] that disturbs the image.

fluctuations in year class strength of 0-group fish.

The life cycle of herring fits well with the scheme of Kawasaki (1980, 1983), while capelin does not. The difference between these two planktivores is probably most pronounced in their spawning areas. Capelin spawns along the southern shores of the Barents Sea, where the adults have to swim through the cod stock in order to cross from the feeding areas to the spawning grounds. In this period, capelin is the dominant prey for cod and mortality of adult capelin is high. If adults should reproduce two seasons in stead of one, they would have to cross this dangerous zone three times in stead of one. If they reproduce only once, they may also allocate more of their body reserves into reproduction, as they do not need to save energy for the return. So this is the trade-off for a coastal spawner having reached the southern shores of the Barents Sea: reproduce once with all available energy (*m* eggs), or save some grams [m (1-q) eggs] with a probability of p^2 to be able to survive two more crossings (assuming all adult mortality risk is associated with the probability p of surviving a crossing through the cod belt). Reproduction twice gives a higher number of offspring than a single reproductive event only if

$$m < m(1-q) + mp^2 \Leftrightarrow q < p^2 \tag{27}$$

According to Eq. 27, semelparous (big-bang) reproduction is optimal at high mortality risk, while low fecundity loss (q) associated with saving energy for survival will favour iteroparity (repeated reproduction). Mortality risk during migration depends both on the sizes of the cod stock and the capelin spawning stocks and fecundity loss q will depend on the body size of the capelin. Hence, the proportion of individuals attempting multiple spawning may vary interannually and with condition. However, if spawning capelin can anticipate the stocks of capelin and cod one year in advance with some accuracy, they are far better qualified than the fisheries biologists.

Timing of habitat shifts

In the process of development from newly hatched larvae to adults, fish undergo substantial changes in anatomy and increases in size by several orders of magnitude. During this growth, the individuals also have to change to new diets, and in most cases growth also leads to changes in preference for place to live. These habitat shifts may be motivated by increased mouth opening that in turn enables the fish to expand its diet, or that larger body reduces risk of predation from one group of predators, while perhaps increasing the risk from other predators. Salvanes & al. (1994) studied the optimal timing for the switch from pelagic to benthic habitat for young cod. They showed that for juveniles, the optimal balance between growth rate (g) and mortality risk (M) was attained by minimising M/g (c.f. Werner & Gilliam 1984, Werner & Hall 1988). Early in life and in the pelagic habitat, mortality rate is extremely high but diminishing (with size and age). As a result there is a period where the ratio M/g is improving, indicating an increased quality of the pelagic habitat for cod larvae. However, the growth rate will eventually be reduced as the cod grows, leading to a decline in profitability of the pelagic habitat later in spring. The benthic habitat is

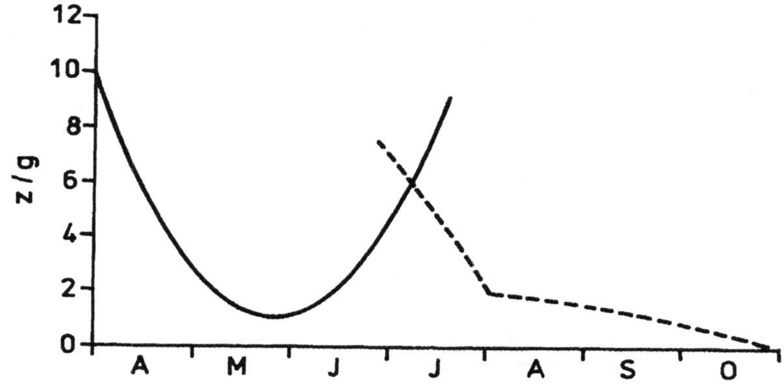

Figure 17. Statistical optimization: optimal timing for settling of cod can be predicted at the intersection between the pelagic and benthic habitat profitability curves. Conditions are initially best in the pelagic habitat (left curve), but through June this habitat is becoming poorer as cod grows, and in the beginning of July conditions are better in the benthic habitat. Redrawn from Salvanes & al. (1994).

extremely poor for newborn cod, mainly due to the food particle size and concentration. However, as cod grow and increases its feeding capacity, a diet shift to larger benthic prey is profitable. Over the same period, the larger codlings becomes more visible to their predators, and shelter becomes important in the habitat quality assessment. Based on data on growth and mortality from different areas, Salvanes & al. (1994) predicted that the settling of cod should take place at midsummer (Fig. 17).

Diet versus depth

In optimal foraging theory, much attention has been put into predicting the diet selection of fish (Townsend & Winfield 1985, Schoener 1987, Gilliam 1990). Giske & Salvanes (1995) showed that the model commonly used is based on the assumption that the time budget of the optimal individual may be divided into separate bouts of feeding and hiding. This is relevant for nearshore and most freshwater fish, but does not apply to open ocean pelagic fish. They then showed that the extra gain in growth achieved by a selective diet in the pelagic could not balance the increased mortality risk associated with staying in so illuminated water that food selection was possible. Rather than be diet-selective, pelagic planktivores should be depth-selective in order to maximise their fitness (see also Box 5). This result may simplify the construction of coupled trophic models of pelagic systems.

Seasonal variation in motivation

Life history theory can be used to predict differences in risk-willingness associated with feeding for juvenile and adult fish. Giske & Aksnes (1992) showed that juveniles that may utilise ingested food to shorten their generation time, benefit more from hazardous feeding than do adults that by feeding only may enhance their egg production. While the optimal trade-off for juveniles is to minimise g/M, adults should minimise lng/M (Fig. 6a).

However, when testing this hypothesis on the mesopelagic planktivore *Maurolicus muelleri* in Masfjorden in winter, they found that adults had so low feeding rate that their growth rate was negative. This is not compatible with any life history model, as an average negative growth rate cannot produce offspring. The negative growth was not associated with poor feeding conditions, as the calculated growth rate of juveniles was very high. (The two age groups separated into two different sound scattering layers, Giske & al. 1990). Adults seemed rather to maximise probability of overwintering survival, which is also observed in dynamic optimisation studies of the life cycle of this planktivore.

Rosland & Giske (1994) modelled the diel vertical distribution of *M. muelleri* during winter by dynamic optimisation. The two age groups were offered the same environment (although visual predation would impact the larger adults far more than juveniles), but the fitness functions. i.e. the relationship between growth over 24 h and fitness gain differed: juveniles were given a high premium for growth, while the reward for adults were almost equal as long as they did not risk short-term starvation. These differences then made the model predict two vertically separated age-groups, very much as observed in the fjord. They also found that the vertical range of each age group, as observed by echo sounder (Giske & al. 1990), corresponded with differences in optimal depth between hungry and satiated individuals. Hence, seasonal variability could explain the overall patterns, while state differences could explain the immediate variability.

Age- and state-dependent motivation

The negative growth rate of adult *Maurolicus muelleri* in winter is an example of the impact of the external environment on motivation and behaviour. From life history theory we also know that motivation will change during the life of an individual, according to the potential fitness value of an action (Giske & Aksnes 1992). This has been studied for the life cycle of a copepod cohort by Fiksen & Giske (1995) by the use of dynamic programming. The environment modelled was constant through time, so the only effect of changes in motivation and behaviour (depth selection) is due to potential changes in growth rate and mortality risk due to the size development of the individual, and to the allocation of assimilated energy to either body growth or egg production. The modelled vertical distribution of the cohort is given in Fig. 18a. The optimal depth of the developmental stages differed, in that larger individuals migrated deeper during daytime to avoid visual predation. However, they overcompensated for predation risk, showing that optimal risk-willingness is reduced as the acquired resources and the probability of survival until first reproduction ("pay day") increases (Fig. 18b).

Also the physiological state (stomach fullness) impacts the short-term motivation and risk-willingness of individuals. Fig. 19 shows that optimal depth is shallower for hungry than for satiated individuals, and that the life-history derived optimum (minimise M/g) applies to neither of them.

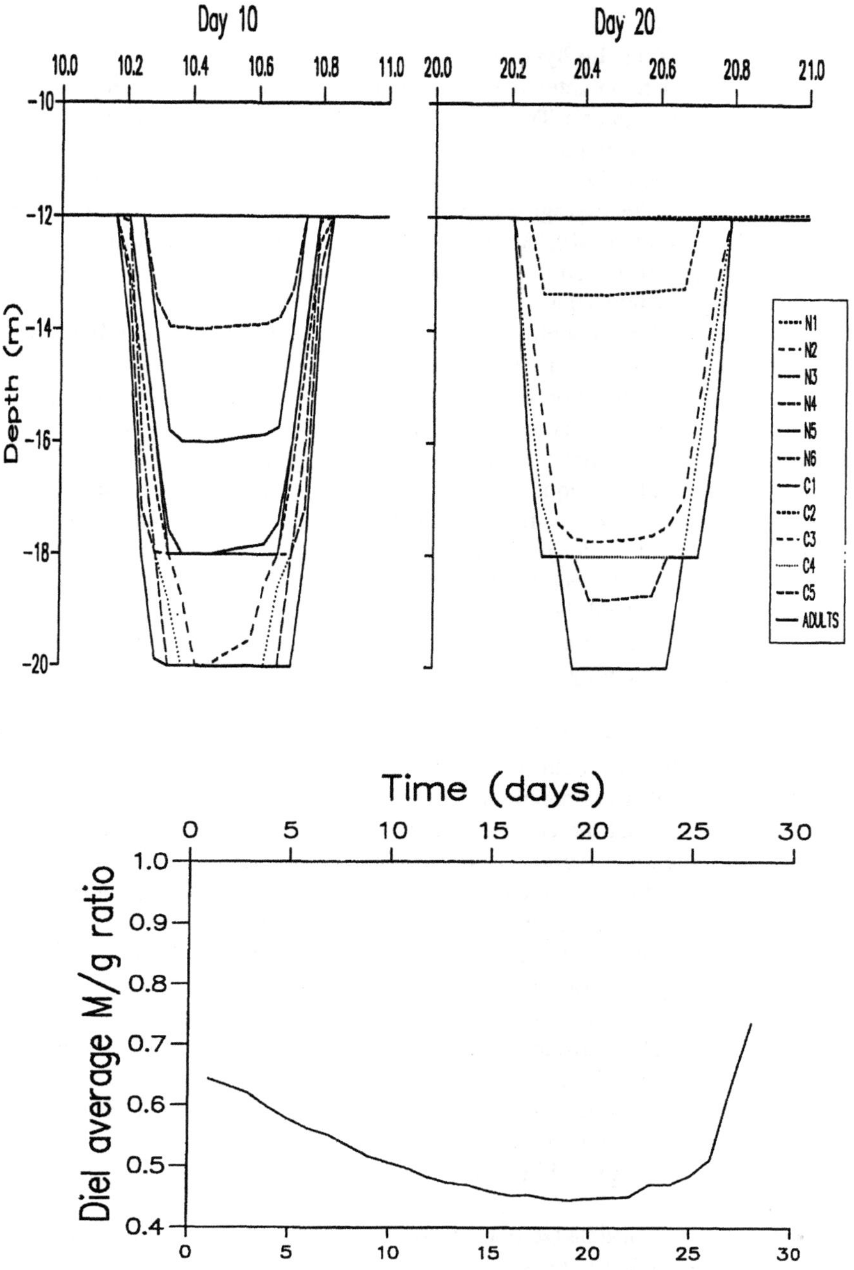

Figure 18. Dynamic optimization: A. The development of the vertical distribution of the copepod cohort. B. Dynamic changes in risk-willingness given as the average mortality risk : growth rate ratio. Redrawn from Fiksen & Giske (1995).

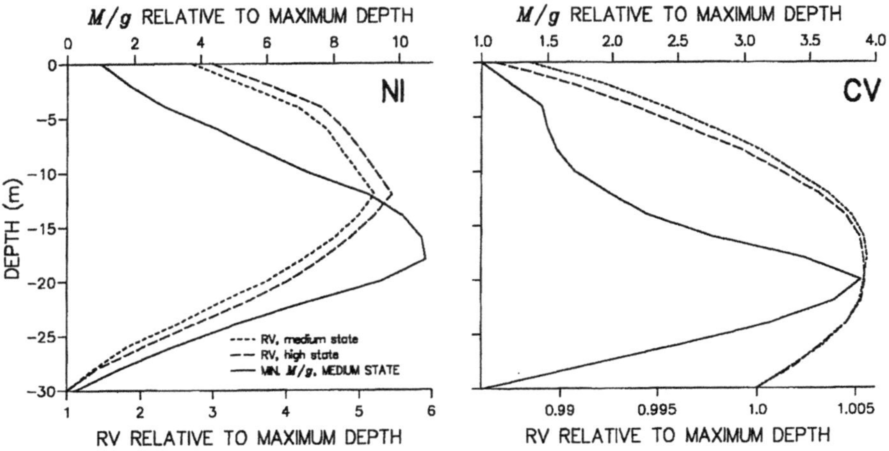

Figure 19. Dynamic optimization: physiological state influences the optimal depth of the copepods. The dotted lines show habitat profitability for NI and CV-individuals of low, medium and high internal state. The solid line depicts habitat profitability as measured by the "minimise M/g" rule, which does not account for internal state. Redrawn from Fiksen (1994).

Effect of density-dependent profitability on copepod distribution

The combination of age-, (size-) and state-dependent motivation of the copepods described above gave a diel range in vertical distribution of 8 meters (11-19 m) with the environmental parameters used. From a constant distribution of the smallest nauplia, diel vertical migration became gradually more emphasised. Later in life the copepods stayed at 17-19 m during daylight hours and at 11-13 m during daytime. Being a dynamic programming approach, the density of copepods was assumed so low that their feeding did not interfere with the dynamics of the food source (phytoplankton). This is a necessary condition for running dynamic programming models over extended time periods. In many practical situation, we are concerned with large populations influencing on their predators and prey, and SDP can then only be used over time scales where these important processes have not had opportunity to affect the environment. These two limitations reduce their value of SDP for practical purposes. Rosland & al. (ms) have therefore studied how density of potential competitors may influence habitat profitability and spatial distributions. An Ideal Free Distribution model was formulated, incorporating 1) dilution of predation risk whenever "competitor" density was high enough to satiate predators, 2) temperature-limited growth rate in absence of competition, but 3) reduction of growth rate when zooplankton community feeding rate exceeded phytoplankton production rate, and finally 4) habitat profitability expressed by the static model of minimise mortality risk per growth rate M/g. This may be a valid assumption during the spring and summer growth season, but does not fit with an overwintering phase (for which dynamic programming would find seasonally adjusted motivations).

There is no diel variability (e.g. in light intensity) in the model, so it does not

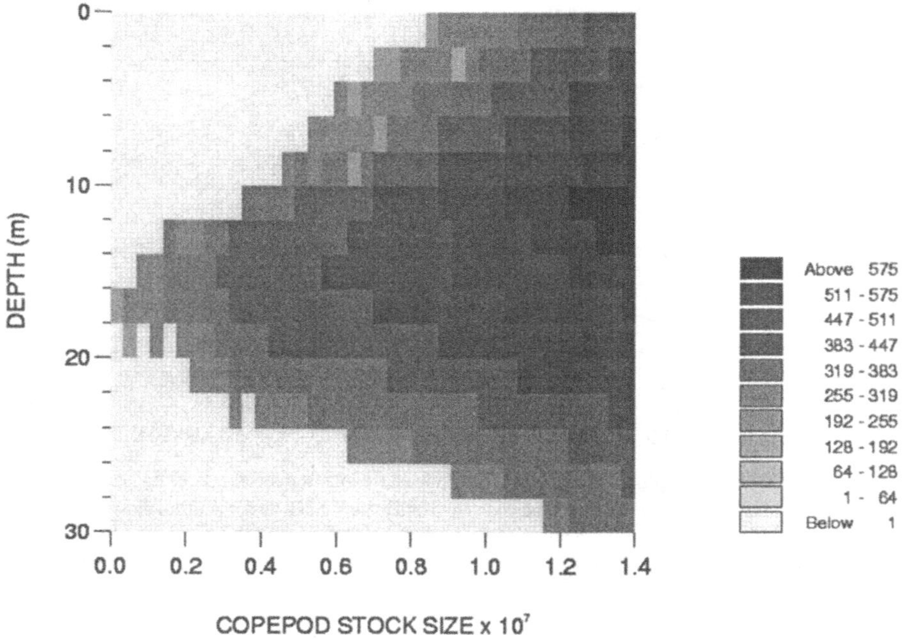

Figure 20. Ideal Free Distribution: optimal vertical distribution depends on the presence of other individuals. As density increases, more depths are utilised, and an increasingly higher fraction of the population seeks shallower depths. Redrawn from Giske & al. 1997.

account for vertical migration. Rather, a optimal daytime distribution is described, dependent on competitor density. With the same environmental data as in Fiksen & Giske (1995), the IFD model finds optimal depth at 12-14 m at very low densities (Fig. 20). With increasing competition, the whole water column (0-30 m) is gradually exploited. As competition increases, the population's vertical distribution will converge towards the profile of the resource input rate. Thus it is evident that optimal distributions should also account for density-dependent factors.

A coupled model for capelin distribution

As a first attempt to make a coupled model of hydrodynamics, zooplankton and fish in the Barents Sea, the model of hydrodynamics and zooplankton (Slagstad 1981a,b, 1987, Slagstad & al. 1989, Støle-Hansen & Slagstad 1991) has been used as input to a dynamic optimisation model of capelin spatial distribution, growth and survival (Fiksen & al. 1995). As there is no interactive link between fish and zooplankton, the model is obviously incomplete. The results from the three-dimensional hydrodynamics and copepod model are summed up to surface-integrated values of temperature and copepod biomass. The growth of capelin is made dependent on swimming costs,

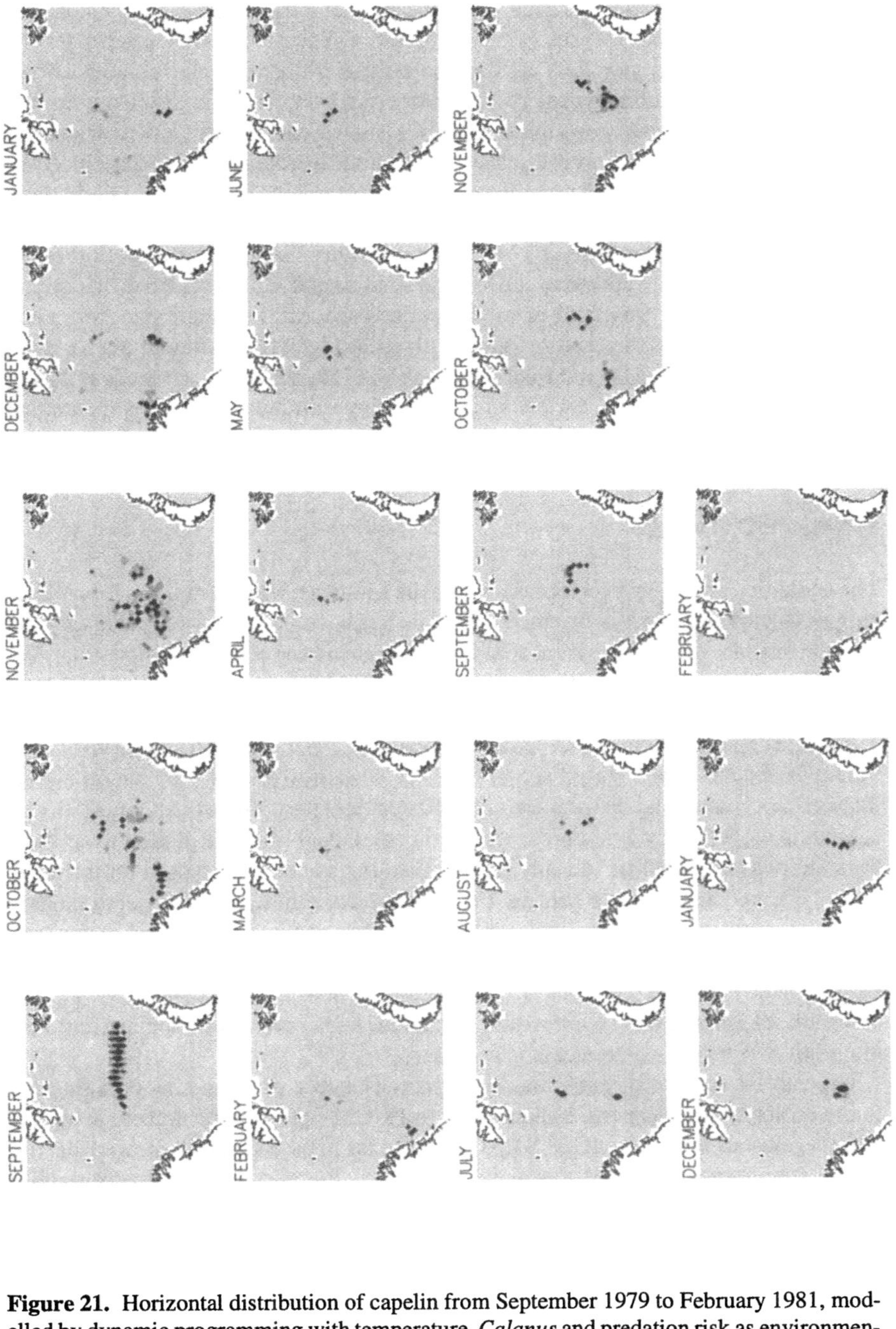

Figure 21. Horizontal distribution of capelin from September 1979 to February 1981, modelled by dynamic programming with temperature, *Calanus* and predation risk as environmental forces. From Fiksen & al. (1995).

temperature and food concentration, using the bioenergetics framework of Kitchell & al. (1977) and Hewett & Johnson (1992). Mortality risk from cod is made dependent on bottom depth and temperature, as we assume that a deeper water column allows capelin some protection against cod predation, and also that the efficiency of cod decreases in colder water. Sea mammals are assumed equally efficient everywhere. The capelin is initialised with the size and spatial distribution as observed on the Institute of Marine Research autumn cruise in 1979, and the behaviour of the capelin is motivated by the fitness premium of spawning along the coast of Finnmark in January-March in either 1980 or 1981. Only individuals larger than 14 g (Forberg & Tjelmeland 1985) are allowed to spawn, and smaller individuals are given a terminal reward in February 1981 (end of model period) according to their size. Modelled monthly dynamics of the capelin stock is given in Fig. 21. The model does not account for the fate of eggs, and traces only the year classes that were present during the autumn cruise in 1979. As this model is quite preliminary and also a theoretical model, no emphasis has been put into tuning output data to observations.

PERSPECTIVES

The ecological modelling performed during the Multispecies program has been rather single-species and process-oriented. And this has been necessary for achieving the long-term goals, given the current state-of-the-art in marine ecological modelling. We have focused on capelin, as a key component in the ecosystem. We will here sketch two ways of incorporating more ecosystem dynamics into the fisheries management.

Fig. 22a depicts a frame for an ecological model of capelin growth and survival. The model is focused on a single species, but ecosystem dynamics are incorporated through stock sizes and distributions of predators and prey. Growth and mortality of capelin is modelled by dynamic programming and ideal free distribution to account for state, season (SDP) and density (IFD). A similar frame may be used for the other actors, e.g. cod and juvenile herring. Fig. 22b depicts a more complex set of models where competition, predation, reproduction and drift of larvae can be run in concert. Here are also indicated direct links between the ecological models and the assessment tools. One of the problems with modelling an ecosystem is that so many processes are interdependent (e.g. feeding of capelin and survival of zooplankton) that integration of model components is necessary.

The current state of physical modelling is reasonably good and encouraging for future ecological applications. Validation of the ECOM-3D model (Blumberg & Mellor 1987) against data from the ICES SKAGEX exercise in the Skagerrak showed that the model reproduced many of the main features of the currents and hydrographic structures (Svendsen et al. 1996). However, it may be difficult to maintain correct descriptions of horizontal and vertical gradients in the density field in models, which may through feed-back have consequences for the modelled current fields. Increased spatial resolution will generally reduce such errors and improve the accuracy of physical models. This increases the model computations, and computing power sets a practical limitation on how far one can go in resolution in coupled physical-chemical-

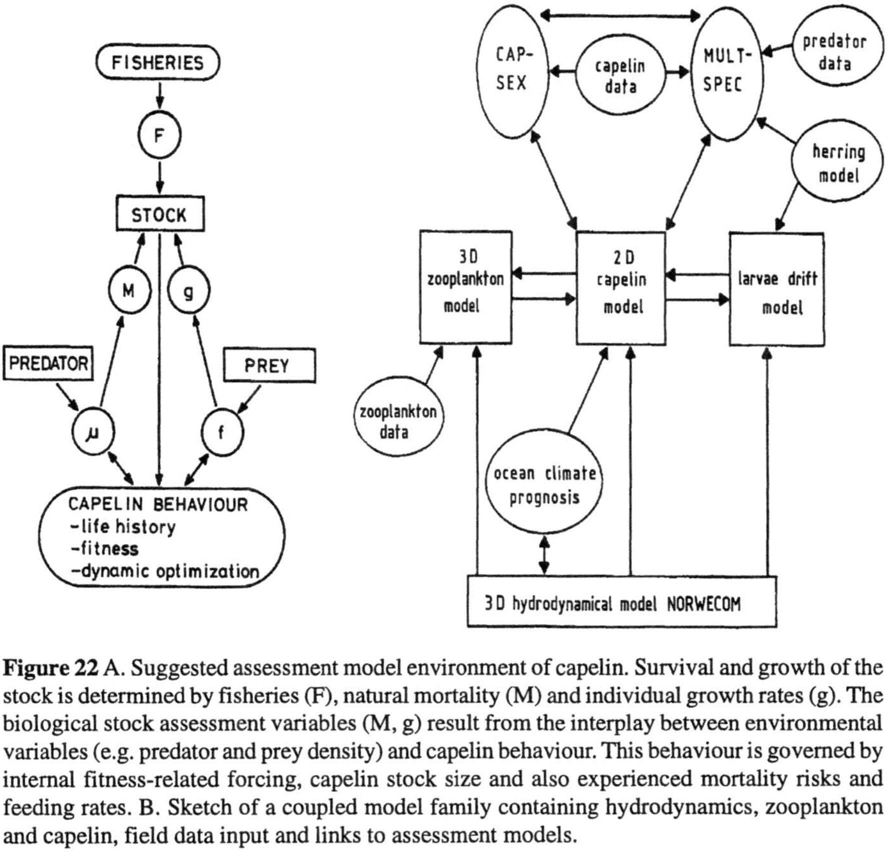

Figure 22 A. Suggested assessment model environment of capelin. Survival and growth of the stock is determined by fisheries (F), natural mortality (M) and individual growth rates (g). The biological stock assessment variables (M, g) result from the interplay between environmental variables (e.g. predator and prey density) and capelin behaviour. This behaviour is governed by internal fitness-related forcing, capelin stock size and also experienced mortality risks and feeding rates. B. Sketch of a coupled model family containing hydrodynamics, zooplankton and capelin, field data input and links to assessment models.

biological models. Major technical challenges lie in parametrisation of smaller-scale processes at a larger model scale, and in nesting finer-scale models within coarser scale.

Physical models have with some success been used to drive models of primary production and advection of zooplankton and fish larvae. Validation against experimental results and comparison with data from the field have shown reasonably good agreement between model and data (Ådlandsvik and Sundby 1994, Slagstad & Stokke 1994, Aksnes et al. 1995, Svendsen et al. 1995). Use of coupled physical-biological models holds great promise for wider use as integral parts of ecological models. Models of advection of fish larvae may provide important information on larval drift routes and geographical recruitment as input to multispecies models (Berntsen & al. 1994, Ådlandsvik & Sundby 1994, Svendsen et al. 1995).

In contrast to fish larvae which typically originate from known and limited spawning grounds, zooplankton spawn over most of their area of distribution (Skjoldal & Melle 1989). Realistic modelling of zooplankton production and advection is therefore dependent on data on the initial fields of zooplankton abundance (Slagstad &

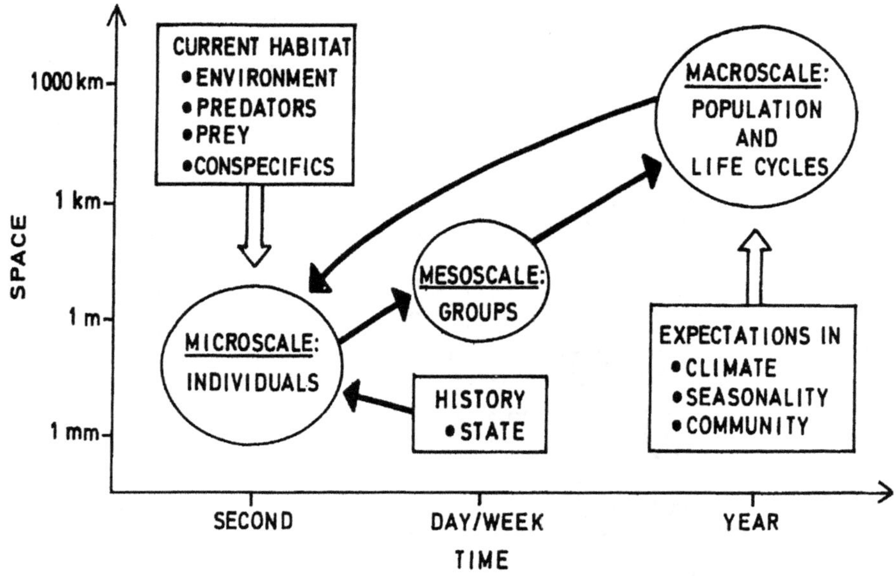

Figure 23. Different methods may be suitable at different scales. Life history theory may give the overall aspects of life in order to understand the match with the physical and biological environment. Dynamic programming is suited to resolve individual differences according to internal state, temporal fluctuations and local environment. Dynamics of schools and shoals may be better studied by density dependent methods like ideal free distribution.

Tande 1996, c.f. see the "Inflow of *Calanus finmarchicus* from the Norwegian Sea" section). A minimum monitoring program of zooplankton in the Barents Sea has been established with the aim to produce data input to models of zooplankton production and advection (Hassel & al. 1993).

From the perspective of ecological modelling of higher trophic levels, the first challenge is to combine the benefits of season- and state-resolution of motivation (dynamic programming) with density dependent habitat profitability (ideal free distribution) into a truly coupled model of several trophic levels (Fig. 23). In this respect, there is also need for more attention to other components in the ecosystem (Fig. 2), both phytoplankton, other groups of zooplankton (krill, amphipods, *Calanus glacialis*) and competitors and predators (herring, cod). Altogether this is a huge task, but on the other side the Barents Sea is a simple ecosystem. There are few ecosystems — if any — of this size and economic interest that are more suited for this enterprise.

However, for the sake of predicting the future, we are still in trouble. If growth and survival of capelin depend on temperature and the distribution of zooplankton, and these depend on hydrodynamics which are heavily influenced by wind force, then we have a time horizon equal to the weather forecast of 4-7 days. Luckily, we are not that bad off. The ocean is not moving as fast as the atmosphere, and changes in the heat content and density structure of large water masses occur quite slowly over some time. The heat and density properties of the ocean have feedback influences on the atmosphere and is likely to act to stabilise the meteorological driving forces. The high degree of regularity and periodicity in ocean climate revealed in oceanographic time series from the Nordic Seas offers some promise that we may be able in the future to forecast the development in ocean climate at time scales of weeks to months or even years. Such predictions may be based on statistical analyses of time series (Ottersen & al. 1994) or better, future insight into the mechanisms and quantitative relationships behind the climatic fluctuations. The current regional GLOBEC program Mare Cognitum on the ecology of the Nordic Seas aims at providing such basic understanding and predictive capability regarding ocean climate (Skjoldal & al. 1993). Also, biological processes are quite slow, except perhaps for growth and survival in the early life phases. There are thus possibilities for creating qualified statements about the near future of plankton and fish stocks in the Barents Sea.

While this chapter has emphasised the large current uncertainties related to the selection of mathematical formulations, this is a problem that may be overcome with time and funding. However, data uncertainty will probably be the dominant source of error in future, mostly due to a strong chaotic element in weather formation. We may not be able to unify all the concepts described above into one model. Rather than constructing one all-encompassing model, we might therefore end up with a series of smaller and different models, each capable to describe one *what-if* question. In combination with statistical techniques to help find the most probable outcomes, the fisheries managers could compare the predictions of several models before making their final subjective decision. In 1995, this seems to us to be as far as ecological modelling can aid fisheries management.

ACKNOWLEDGEMENTS

We thank Øyvind Fiksen, Rune Rosland and Anne Gro Salvanes for discussions and improvements of the text. This work has been supported by the Research Council of Norway, mainly through the Multispecies Program.

REFERENCES

Abrahams, M.V. 1986. Patch choice under perceptual constraints: a cause for departures from an ideal free distribution. Behav. Ecol. Sociobiol. 19: 409-415.

Ådlandsvik, B. & H. Loeng 1991. A study of the climatic system in the Barents Sea. Polar Res. 10: 45-49.

Ådlandsvik, B. & S. Sundby 1994. Modelling the transport of cod larvae from the Lofoten area. ICES mar. Sci. Symp. 198: 379-392.

Aksnes, D.L., J. Aure, S. Kaartvedt, T. Magnesen & J. Richard 1989. Significance of advection for the carrying capacities of fjord populations. Mar. Ecol. Prog. Ser. 50: 263-274.

Aksnes, D.L. & J. Giske 1990. Habitat profitability in pelagic environments. Mar. Ecol.

Prog. Ser. 64: 209-215.

Aksnes, D.L. & J. Giske 1993. A theoretical model of aquatic visual feeding. Ecol. Model. 67: 233-250.

Aksnes, D.L. & U. Lie 1990. A coupled physical-biological pelagic model of a shallow sill fjord. Estu. Coast. Shelf Sci. 31: 459-486.

Aksnes, D.L., K.B. Ulvestad, B.M. Baliño, J. Berntsen, J.K. Egge & E. Svendsen 1995. Ecological modelling in coastal waters: towards predictive physical-chemical-biologicl simulation models. Ophelia 41: 5-36.

Alvheim, O. 1985. Investigations on capelin larvae off northern Norway and in the Barents Sea in 1981-84. Pp. 171-183 in: Gjøsæter, H. (ed.). Proceedings of the Soviet-Norwegian symposium on the Barents Sea capelin. Institute of Marine Research, Bergen, Norway.

Armstrong, M.J. & P.A. Shelton 1990. Clupeoid life-history styles in variable environments. Environ. Biol. Fishes 28: 77-85.

Bailey, K.M. & E.D. Houde 1989. Predation on eggs and larvae of marine fishes and the recruitment problem. Adv. mar. Biol. 25, 1-83.

Balchen J.G. 1976a. Principles of migration in fishes. SINTEF Rep. STF48 A76045, Trondheim, Norway. 33 pp.

Balchen, J.G. 1976b. Modelling of the biological state of fishes. SINTEF Rep. STF48 A76023, Trondheim, Norway. 25 pp.

Balchen, J.G. 1980. Modeling and identification of marine ecological systems with applications in the management of fish resources and planning of fisheries operations. Model. Ident. Control. 1: 67-68.

Baretta, J.W., W. Ebenhöh & P. Ruardij 1995. The European Regional Seas Ecosystem Model, a complex marine ecosystem model. Neth. J. Sea Res. 33: 233-246.

Berntsen, H., Z. Kowalik, S. Sælid & K. Sørli 1981. Efficient numerical simulation of ocean hydrodynamics by a splitting procedure. Model. Ident. Control. 2: 181-199.

Berntsen, J., D.W. Skagen & E. Svendsen 1994. Modeling the transport of particles in the North Sea with reference to sandeel larvae. Fish. Oceanogr. 3: 81-89.

Bird, R.E. 1984. A simple, solar spectral model for directnormal and diffuse horizontal irradiance. Solar Energy 32: 461-471.

Bjørndal, T. 1988. The optimal management of North Sea herring. J. Environ. Econ. 15: 9-29.

Bjørndal, T., J. Conrad & K.G. Salvanes 1993. Stock size, harvesting cost, and the potential for extinction. Land Econ. 69: 156-167.

Blindheim, J. & H.R. Skjoldal 1993. Effects of climatic changes on the biomass yield of the Barents Sea, Norwegian Sea, and West Greenland large marine ecosystems. Pp. 185-198 in: K. Sherman, L.M. Alexander & B.D. Gold, (eds.). Large Marine Ecosystems: stress, mitigation and sustainability. AAAS Publ. 92-39S. AAAS Publications, Washington DC, USA.

Blumberg, A.F. & G.L. Mellor 1987. A description of a three-dimensional coastal ocean circulation model. In: Heaps, N. (ed.). Three-dimensional coastal ocean models, Vol 4. American Geophysical Union, Washington DC. USA.

Bogstad, B., S. Tjelmeland, T. Tjelta & Ø. Ulltang 1992. Description of a multispecies

model for the Barents Sea (MULTSPEC) and a study of its sensitivity to assumptions on food preferences and stock sizes of minke whales and harp seals. Int. Whal. Commn. sci. Comm. 09: 47 pp.
Calow, P. & C.R. Townsend 1981. Energetics, ecology and evolution. Pp. 3-19 in: Townsend, C.R. & P. Calow (eds.). Physiological ecology. An evolutionary approach to resource use. Blackwell, Oxford.
Charles, A.T. 1983. Optimal fisheries investment: comparative dynamics for a deterministic seasonal fishery. Can. J. Fish. Aquat. Sci. 40: 2069-2079.
Clark, C.W. 1976. Mathematical bioeconomics: the optimal management of renewable resources. Wiley, NY.
Clark, C.W. & D.A. Levy 1988. Diel vertical migration by juvenile sockeye salmon and the antipredation window. Am. Nat. 131: 271-290.
Clark, C.W. & M. Mangel 1986. The evolutionary advantages of group foraging. Theor. Pop. Biol. 12: 119-129.
Colebrook, J. M. 1985. Sea surface temperature and zooplankton, North Sea 1948 to 1983. J. Cons. int. Explor. Mer. 42: 179-185.
Devold, F. 1963. The life history of the Atlanto-Scandian herring. Rapp. P.-v. Réun. cons. perm. int. Explor. Mer 154: 98-108.
Dickson, R. R., J. Meincke, S.A. Malmberg & A.L. Lee 1988. The "Great Salinity Anomaly" in the North Atlantic 1968-1982. Prog. Oceanogr. 20: 103-151.
Dragesund, O. & J. Gjøsæter 1988. The Barents Sea. Pp. 339-361 in Postma, H. & J.J. Zijlstra (eds). Continental Shelves, Elsevier.
Dragesund, O., J. Hamre and Ø. Ulltang 1980. Biology and population dynamics of the Norwegian spring-spawning herring. Rapp. P.-v. Réun. Cons. int. Explor. Mer 177: 43-71.
Emlen, J.M. 1966. The role of time and energy in food preference. Am. Nat. 100: 611-617.
Euler, L. 1760. Recherches générales sur la mortalité: la multiplication du genre humain. Mem. Acad. Sci. Berlin 16: 144-164.
Fiksen, Ø. 1994. Vertical distribution and population dynamics of copepods by dynamic optimization. Cand. scient. thesis, University of Bergen, Department of Fisheries and Marine Biology, Bergen, Norway. 130 pp.
Fiksen, Ø. & J. Giske 1995. Vertical distribution and population dynamics of copepods by dynamic optimization. ICES J. mar. Sci. 52: 483-503.
Fiksen, Ø., J. Giske and D. Slagstad 1995. A spatially explicit fitness-based model of capelin migrations in the Barents Sea. Fish. Oceanogr. 4: 193-208.
Flåten, O. 1988. *The economics of multispecies harvesting theory and application to the Barents Sea fisheries*. Springer Verlag, Berlin.
Forberg, K. & S. Tjelmeland 1985. Maturity studies of Barents Sea capelin. Variations in length at maturity for female capelin. Pp. 213-222 in: Gjøsæter, H. (ed.). Proceedings of the Soviet-Norwegian symposium on the Barents Sea capelin. Institute of Marine Research, Bergen, Norway.
Fossum, P. 1992. The recovery of the Barents Sea capelin *Mallotus villosus* from a larval point of view. ICES J. mar. Sci. 49: 237-243.
Fretwell, S.D. 1972. *Populations in a seasonal environment.* Princeton University

Press, Princeton, New Jersey.
Fretwell, S.D. & H.J. Lucas Jr. 1970. On territorial behavior and other factors influencing habitat distributions in birds. 1. Theoretical development. Acta Biotheo. 19: 16-36.
Gilliam, J.F. 1982. Habitat use and competitive bottlenecks in size-structured fish populations. Ph. D. dissertation, Michigan State University, East Lansing. 107 pp.
Gilliam, J. F. 1990. Hunting by the hunted: optimal prey selection by foragers under predation hazard. Pp. 797-819. in: Hughes, R. N. (ed.). Behavioural mechanisms of food selection. NATO ASI Series Vol. G 20, Berlin: Springer-Verlag.
Giske, J. & D.L. Aksnes 1991. Vertikalfordelinger: hvor dypt befinner fisk og plankton seg? Naturen 115: 195-202.
Giske, J. & D.L. Aksnes 1992. Ontogeny, season and trade-offs: vertical distribution of the mesopelagic fish *Maurolicus muelleri*. Sarsia 77: 253-261.
Giske, J., D.L. Aksnes, B.M. Baliño, S. Kaartvedt, U. Lie, J.T. Nordeide, A.G.V. Salvanes, S.M. Wakili & A. Aadnesen 1990. Vertical distribution and trophic interactions of zooplankton and fish in Masfjorden, Norway. Sarsia 75: 65-82.
Giske, J., D.L. Aksnes & Ø. Fiksen 1994. Visual predators, environmental variables and zooplankton mortality risk. Vie Milieu 44: 1-9.
Giske, J., D.L. Aksnes & B. Førland 1993. Variable generation times and Darwinian fitness measures. Evol. Ecol. 7: 233-239.
Giske, J., D.L. Aksnes, U. Lie & S.M. Wakili 1991. Computer simulation of pelagic production in Masfjorden, Western Norway, and its consequences for production of released 0-group cod. ICES mar. Sci. Symp. 192: 161-175.
Giske, J., R. Rosland, J. Berntsen and Ø. Fiksen 1997. Ideal free distribution of copepods under predation risk. Ecol. Mod. 95: 45-59.
Giske, J. & A.G.V. Salvanes 1995. Why pelagic predators should be unselective feeders. J. theor. Biol. 173: 41-50.
Giske, J., H.R. Skjoldal & D.L. Aksnes 1992a. A conceptual model of distribution of capelin in the Barents Sea. Sarsia 77: 147-156.
Giske, J., H.R. Skjoldal & D.L. Aksnes 1992b. Modellering av fordeling, vekst og overlevelse av fisk i Barentshavet (Modelling distribution, growth and survival of fish in the Barents Sea). Fisken Hav. 1992(1): 20 pp.
Gjøsæter, H. 1985. Growth of the Barents Sea capelin of the yearclasses 1975-1981. Pp. 193-212 in: Gjøsæter, H. (ed.). Proceedings of the Soviet-Norwegian symposium on the Barents Sea capelin. Institute of Marine Research, Bergen.
Gjøsæter, H. & H. Loeng 1987. Growth of the Barents Sea capelin, *Mallotus villosus* in relation to climate. Environ. Biol. Fish. 20: 293-300.
Godin, J.G. J. & S.A. Smith. 1988. A fitness cost of foraging in the guppy. Nature 333: 69-71.
Godø, O.R. & H.R. Skjoldal 1991. Fluctuations in abundance of the Barents Sea cod related with environmental and ecological changes. Cod Symposium, Sept. 1991. Serial No. N2010. NAFO SCR Doc. 91/117.
Gray, R.D. & M. Kennedy 1994. Perceptual constraints on optimal foraging: A reason

for departures from the ideal free distribution? Anim. Behav. 47: 469-471.
Gulland, J.A. 1965. Estimation of mortality rates. Annex to Arctic Fisheries Working Group Report (meeting in Hamburg, January 1965). ICES ĆM 1965. Doc. No. 3: 9 pp.
Håkanson, L. 1995. Optimal size of predictive models. Ecol. Mod. 78: 195-204.
Hamre, J. 1985. Assessment and management of Barents Sea capelin. Pp. 213-222 in: Gjøsæter, H. (ed.). Proceedings of the Soviet-Norwegian symposium on the Barents Sea capelin. Institute of Marine Research, Bergen.
Hamre, J. 1991. Interrelation between environmental changes and fluctuating fish populations in the Barents Sea. Pp. 259-270 in Kawasaki, T., S. Tanaka, Y. Toba & A. Taniguchi (eds.). Long-term variability of pelagic fish populations and their environment. Pergamon Press, Tokyo.
Hanneson, R. 1993. *Bioeconomic analyses of fisheries*. Blackwell Scientific Publications, London.
Harms, I. 1994. Numerische Modellstudie zur winterlichen Wassermassenformation in der Berentssee. Ph. D. thesis. Institut für Meereskunde. Hamburg.
Hassel, A., H.R. Skjoldal & M. Johannessen 1993. Zooplankton in the Barents Sea: Monitoring as basis for modelling. Multspec-News, Autumn 1993, Institute of Marine Research, Bergen. 3 pp.
Helland-Hansen, B. & F. Nansen 1909. The Norwegian Sea: Its physical oceanography based upon the Norwegian researches 1900-1904. FiskDir. Skr. Ser. HavUnders. 2: 1-360.
Hewett, S.W. & B.J. Johnson 1992. An upgrade of a generalized bioenergetics model of fish growth for microcomputers. University of Wisconsin, Wisconsin Sea Grant College Program, Sea Grant Techn. Rep., WIS-SG-92-250, Madison.
Hjort, J. 1902. Fiskeri og Hvalfangst. Aarsberetning vedk. Norges fiskerier. Bergen.
Hjort, J. 1914. Fluctuation in the great fisheries in northern Europe viewed in the light of biological research. Rapp. P.-v. Réun. Cons. int. Explor. Mer 20: 1-228.
Holland, J.H. 1975. *Adaptation in natural and artificial systems*. University of Michigan Press, Ann Arbor.
Holland, J.H. 1992. Genetic algorithms. Scientific American, July 1992, 44-50.
Houston, A., C. Clark, J. McNamara & M. Mangel 1988. Dynamic models in behavioural and evolutionary ecology. Nature (London) 332: 29-34.
Jakobsen, P. J. & G. H. Johnsen 1988. Size-specific protection against predation by fish in swarming waterfleas, *Bosmina longispina*. Anim. Behav. 36: 986-990.
Jensen, A.L. 1991. Simulation of fish population responses to exploitation. Ecol. Model. 55: 203-218.
Jørgensen, T. 1988. Long-term changes in age at sexual maturity of the Northeast Artic cod (*Gadus morhua* L.). ICES CM 1988/G:42, 32 pp.
Kawasaki, T. 1980. Fundamental relations among the selections of life history in the marine teleosts. Bull. Jpn. Soc. Sci. Fish. 46: 289-293.
Kawasaki, T. 1983. Why do some pelagic fishes have wide variations in their numbers? Biological basis of fluctuation from the viewpoint of evolutionary ecology. FAO Fish. Rep. 291: 1065-1080.

Kirk, J.T.O. 1984. *Light and photosynthesis in aquatic ecosystems*. Cambridge University press, Cambridge. 416 pp.

Kitchell, J.F., D.J. Stewart & D. Weininger 1977. Application of a bioenergetics model to yellow perch (*Perca flavescens*) and walleye (*Stizostedion vitreum*). J. Fish. Res. Board Can. 37: 2159-2168.

Law, R. & D.R. Grey 1989. Evolution of yields from populations with age-specific cropping. Evol. Ecol. 3: 343-359.

Levitus, S. (ed.) 1982. Climatological atlas of the world ocean. N.O.A.A. Professional Paper, Rockville, Maryland. 173 pp.

Loeng, H. 1989a. Ecological features of the Barents Sea. Pp. 327-365 in: Rey, L. & V. Alexander (eds). Proc. sixth Conf. Comité Arctique Intern.: 13-15 May 1985. E.J. Brill.

Loeng, H. 1989b. The influence of temperature on some fish population parameters in the Barents Sea. J. Northw. Atl. Fish. Sci. 9: 103-113.

Loeng, H., J. Blindheim, B. Ådlandsvik & G. Ottersen 1992. Climatic variability in the Norwegian and Barents Seas. ICES mar. Sci. Symp. 195: 52-61.

Lotka, A.J. 1907. Studies on the norm of growth of material aggregates. Am. J. Sci. 24: 199-216, 375-376.

MacArthur, R.H. & E.R. Pianka 1966. On optimal use of a patchy environment. Am. Nat. 100: 603-609.

MacCall, A.D. 1990. *Dynamic geography of marine fish populations.* Books in recruitment fishery oceanography, Washington Sea Grant Program, Seattle. 153 pp.

Magnússon, K.G. 1995. An overview of the multispecies VPA - theory and applications. Rev. Fish Biol. Fisheries. 5: 195-212.

Magurran, A.E., W.J. Oulton & T.J. Pitcher 1985. Vigilant behaviour and shoal size in minnows. Zeitschr. Tierpsychol. 67: 167-178.

Mangel, M. 1990. Resource divisibility, predation and group formation. Anim. Behav. 39: 1163-1172.

Mangel, M. & C.W. Clark 1986. Towards a unified foraging theory. Ecology 67: 1127-1138.

Mangel, M. & C.W. Clark 1988. *Dynamic modeling in behavioral ecology*. Princeton University Press, Princeton, New Jersey. 308 pp.

McClimans, T. A. 1993. An algorithm for computing monthly averaged inflow of Atlantic water to the Norwegian Sea. SINTEF NHL Rep. STF60 A93009.

Midttun, L. 1985. Formation of dense bottom water in the Barents Sea. Deep-Sea Res. 32: 1233-1241.

Midttun, L. 1989. Climatic fluctuations in the Barents Sea. Rapp. P.-v. Réun. Cons. int. Explor. Mer. 188: 23-35.

Milinski, M. 1977. Do all members of the swarm suffer the same predation? Zeitschr. Tierpsychol. 45: 373-388.

Milinski, M. 1979. Evolutionarily stable feeding strategies in sticklebacks. Zeitschr. Tierpsychol. 51: 36-40.

Milinski, M. 1984. A predator's cost of overcoming the confusion effect of swarming

prey. Anim. Behav. 32: 1157-1162.
Milinski, M. 1985. Risk of predation taken by parasitised sticklebacks under competition for food. Behav. 93: 203-216.
Milinski, M. 1986. Constraints placed by predators on feeding behaviour. - Pp. 236-252 in: Pitcher, T.J. (ed.). The behaviour of teleost fishes. Croom Helm, London.
Miller, T. J., L.B. Crowder, J.A. Rice & E.A. Marschall 1988. Larval size and recruitment mechanisms in fishes: towards a conceptual framework. Can. J. Fish. Aquat. Sci. 45: 1657-1670.
Murphy, G. I. 1968. Patterns in life history and the environment. Am. Nat. 102: 391-403.
Nakken, O. & A. Raknes 1987. The distribution and growth of north-east Arctic Cod in relation to bottom temperatures in the Barents Sea, 1978-1984. Fish. Res. 5: 243-252.
Øiestad, V. 1990. Konsekvenser av klimaendringer for fiskeri- og havbruksnæringen. Fisken Hav. 2/1990.
Østvedt, O.-J. 1955. Zooplankton investigations from weather ship M in the Norwegian Sea, 1948-49. Hvalrådets Skrifter, Sci. Resul. mar. Biol. Res. 40: 93 pp.
Ottersen, G., Ådlandsvik, B. and Loeng, H. 1994. Statistical modelling of temperature variability in the Barents Sea. ICES C.M. S:2. Norway.
Ozhigin, V.K. & G.I. Luka 1985. Some peculiarities of capelin migrations depending on thermal conditions in the Barents Sea. Pp. 135-147 in: Gjøsæter, H. (ed.). Proceedings of the Soviet-Norwegian symposium on the Barents Sea capelin. Institute of Marine Research, Bergen.
Ozhigin, V.K. & N.G. Ushakov 1985. The effect of the thermal conditions of the sea and atmospheric circulation on the distribution of the Barents Sea capelin feeding areas. Pp. 149-156 in: Gjøsæter, H. (ed.). Proceedings of the Soviet-Norwegian symposium on the Barents Sea capelin. Institute of Marine Research, Bergen.
Parkin, D.T. 1979. *An introduction to evolutionary genetics*. Edward Arnold, London. 223 pp.
Parsons, T. R., M. Takahashi & B. Hardgrave 1983. *Biological oceanographic processes*. Pergamon Press. Oxford. 332 pp.
Pedersen, G., K. Tande & G.O. Ottesen 1995. Why does *Calanus finmarchicus* stay in surface waters during the overwintering period in high latitudes? ICES J. mar. Sci. 52: 523-531.
Phillipson, J. 1981. Bioenergetic options and phylogeny. Pp. 20-45 in: Townsend, C.R. & P. Calow (eds.). Physiological ecology. An evolutionary approach to resource use. Blackwell, Oxford.
Pope, J. 1972. An investigation of the accuracy of virtual population analysis using cohort analysis. Res. Bul. int. Comm. Northw. Atl. Fish. 9: 65-74.
Pope, J. 1979. Population dynamics and management: current status and future trends. Invest. Pesq. 43: 199-211.
Rey, F., H.R. Skjoldal & D. Slagstad 1987. Primary production in relation to climatic changes in the Barents Sea. Pp. 29-46 in: Loeng, H. (ed.). The effect of oceanographic conditions on distribution and population dynamics of commercial fish stocks in the Barents Sea. Proceedings of the third Soviet-Norwegian Sympo-

sium, Murmansk, 26-28 May 1986. Institute of Marine Research, Bergen.
Rosland, R. & J. Giske 1994. A dynamic optimization model of the diel vertical distribution of a pelagic planktivorous fish. Prog. Oceanog. 34: 1-43.
Sakshaug, E., A. Bjørge, B. Gulliksen, H. Loeng & F. Mehlum (eds.) 1992. *Økosystem Barentshav*. PRO MARE, Norwegian Research Program for Marine Arctic Ecology, Norges Allmennvitenskapelige Forskningsråd, Oslo. 304 pp.
Sakshaug, E., A. Bjørge, B. Gulliksen, H. Loeng & F. Mehlum 1994. Structure, biomass distribution, and energetics of the pelagic ecosystem in the Barents Sea: A synopsis. Polar Biol. 14: 405-411.
Sakshaug, E., C.C.E. Hopkins & N.A. Øritsland (Eds.) 1991. Proceedings of the *Pro Mare* symposium on Polar Marine Ecology, Trondheim, Norway, 12-16 May 1990. Polar Res. 10 (1-2), 662 pp.
Sakshaug, E., F. Rey & D. Slagstad 1995. Wind forcing of marine primary production in the Northern atmospheric low-pressure belt. MARE NOR Proceedings.
Sakshaug, E. & H.R. Skjoldal 1989. Life at the ice edge. Ambio 18: 60-67.
Salvanes, A.G.V., D.L. Aksnes, J.H. Fosså & J. Giske 1995. Simulated carrying capacities of fish in Norwegian fjords. Fish. Oceanog. 4: 17-32.
Salvanes, A.G.V., D.L. Aksnes & J. Giske 1992. Ecosystem model for evaluating potential cod production in a west Norwegian fjord. Mar. Ecol. Prog. Ser. 90: 9-22.
Salvanes, A.G.V., J. Giske & J.T. Nordeide 1994. A life-history approach to habitat shifts for coastal cod (*Gadus morhua* L.). Aquac. Fish. Man. 25(Suppl. 1): 215-228.
Sætersdal, G. & H. Loeng 1987. Ecological adaptation of reproduction in Northeast Arctic cod. Fish. Res.5: 253-270.
Schoener, T.W. 1987. A brief history of optimal foraging theory. Pp. 5-67 in: Kamil, A.C., J.R. Krebs & H.R. Pulliam (eds.). Foraging behaviour. Plenum Press, New York.
Sherman, K., L.M. Alexander & B.D. Gold (eds.) 1993. Large Marine Ecosystems: stress, mitigation and sustainability. AAAS Publ. 92-39S. AAAS Publications, Washington DC, USA. 376 pp.
Sibly, R.M. & P. Calow 1986. Why breeding earlier is always worthwhile. J. theor. Biol. 123: 311-319.
Sinclair, M. 1988. Marine populations. An essay on population regulation and speciation. - Books in recruitment fishery oceanography, Washington Sea Grant. University of Washington Press, Seattle. 252 pp.
Sinclair, M. & P. Solemdal. 1988. The development of "population thinking" in fisheries biology between 1878 and 1930. Aqu. liv. reso. 1: 189-213.
Skjoldal, H.R. 1990. Management of marine living resources in a changing ocean climate. Pp. 1-17 in: Papers presented on the session "Research on natural resources management" of the Conference "Sustainable development, science and policy", Bergen 8-12 May 1990. Bergen Foundation of Science Rep. No. BFS A90005. Bergen, Norway.
Skjoldal, H.R., A. Hassel, F. Rey & H. Loeng 1987. Spring phytoplankton development and zooplankton reproduction in the central Barents Sea in the period

1979-1984. Pp. 59-89 in: Loeng, H. (ed.). The effect of oceanographic conditions on distribution and population dynamics of commercial fish stocks in the Barents Sea. Proceedings of the third Soviet-Norwegian Symposium, Murmansk, 26-28 May 1986. Institute of Marine Research, Bergen.
Skjoldal, H.R. & W. Melle 1989. Nekton and plankton: Some comparative aspects of larval ecology and recruitment processes. Rapp. P.-v. Reun. Cons. int. Explor. Mer 191: 330-338.
Skjoldal, H.R., T.T. Noji, J. Giske, J.H. Fosså, J. Blindheim & S. Sundby 1993. Mare Cognitum. Science plan for research on marine ecology of the Nordic Seas (Greenland, Norwegian, Iceland Seas) 1993-2000. Institute of Marine Research, Bergen, Norway. 162 pp.
Skjoldal, H.R. & F. Rey 1989. Pelagic production and variability in the Barents Sea ecosystem. Pp. 241-286 in: Sherman, K. & L.M. Alexander, (eds.). Biomass yields and geography of large marine ecosystems. AAAS Selected symposium 111. American Association for the Advancement of Science, Warshington, USA.
Slagstad, D. 1981a. The copepod model adapted to an Arctic/Boreal environment. Sensitivity analysis. SINTEF rep. STF48 F81056.
Slagstad, D. 1981b. Modeling and simulation of physiology and population dynamics of copepods. Effects of physical and biological parameters. Model. Ident. Control 2: 119-162.
Slagstad, D. 1987. A 4-Dimensional physical model of the Barents Sea. SINTEF Rep., Trondheim STF48 F87013: 1-34
Slagstad, D & S. Stokke 1994. Simulering av strømfelt, hydrografi, isdekke og primærproduksjon i det Nordlige Barentshav. Fisken Hav. 9: 47 pp.
Slagstad, D. & K. Støle-Hansen 1991. Dynamics of plankton growth in the Barents Sea. Model studies. Polar Res. 10: 173-186.
Slagstad, D., K. Støle-Hansen & H. Loeng 1989. Density driven currents in the Barents Sea calculated by a numerical model. Model. Ident. Control 11: 181-190
Slagstad, D. & K. Tande 1985. Factors affecting the vertical migration behaviour of copepods. A mathematical model. SINTEF rep. STF48 F85021, Dec. 1985.
Slagstad, D. & K. Tande 1990. Growth and production dynamics of Calanus glacialis in an arctic pelagic food web. Mar. Ecol. Prog. Ser. 63: 189-199.
Slagstad, D. & K. Tande 1996. The importance of seasonal vertical migration in across shelf transport of *Calanus finmarchicus*. Ophelia 44: 189-205.
Slagstad, D., T. Westgård, K. Olsen, S. Sælid & J.G. Balchen 1975. Mathematical modelling of population, quality, migration and distribution of important species of fish in an ocean. (In Norwegian). SINTEF Rap. STF48 A75050, Trondheim, Norway. 119 pp.
Stearns, S.C. & R.E. Crandall 1984. Plasticity for age and size at sexual maturity: a life-history response to unavoidable stress. Pp. 13-29. in: Potts, G. W., Wootton, R. J. (eds.). Fish reproduction, strategies and tactics. Academic Press. London.
Stephens, D.W. & J.R. Krebs 1986. *Foraging theory*. Princeton University Press, Princeton, New Jersey.
Støle-Hansen, K., & D. Slagstad 1991. Simulations of currents, ice-melting and verti-

cal mixing in the Barents Sea using a 3-D baroclinic model. Polar Res. 10: 33-44.
Sutherland, W.J. 1990. Evolution and fisheries. Nature (London) 344: 814-815.
Svendsen, E., J. Berntsen, M.D. Skogen, B. Ådlandsvik & E. Martinsen 1996. Model simulation of the Skagerrak circulation and hydrography during Skagex. J. Mar. Syst. 8: 219-236.
Svendsen, E., P. Fossum, M.D. Skogen, G. Eriksrød, H. Bjørke, K. Nedreaas & A. Johannessen 1995. Variability of the drift patterns of spring spawned herring larvae and the transport of water along the Norwegian shelf. ICES C.M./Q: 25. 31 pp.
Tande, K. & D. Slagstad 1992. Regional and interannual variations in biomass and productivity of the marine copepod, Calanus finmarchicus, in subarctic environments. Oceanologia Acta., 15: 309-321.
Tjelmeland, S. & B. Bogstad 1997. Biological modelling. Pp. 69-91 in: Rødseth, T. (ed.). *Models for multi-species management.* Springer Verlag. Berlin.
Townsend, C.R. & I.J. Winfield 1985. The application of optimal foraging theory to feeding behaviour in fish. Pp. 67-98 in: Tytler, P. & P. Calow (eds.). Fish energetics - new perspectives. Baltimore: The Johns Hopkins University Press.
Tyler, J.A. & D.P. Clapp 1995. Perceptual constraints on stream fish habitat selection: effects of food availability and water velocity. Ecol. Fres. Fish 4: 9-16.
Tyler, J.A. & J.F. Gilliam 1995. Ideal free distributions of stream fish: a model and test with minnows, *Rhinicthys atratulus*. Ecology 76: 580-592.
Ulltang, Ø. 1995. Multispecies modelling and management with reference to the Institute of Marine Research's multispecies model for the Barents Sea. Pp. 659-670 in: A.S. Blix, L. Walløe and Ø. Ulltang (eds.): Whales, seals, fish and man. Elsevier Science.
Ushakov, N.G. & V.K. Ozhigin 1987. The abundances of year-classes of the Barents Sea capelin and the peculiarities of the distribution of the young in relation to hydrographic conditions. Pp. 159-167 in: Loeng, H. (ed.). The effect of oceanographic conditions on distribution and population dynamics of commercial fish stocks in the Barents Sea. Proceedings of the third Soviet-Norwegian Symposium, Murmansk, 26-28 May 1986. Institute of Marine Research, Bergen.
Utne, A.C.W. & D.L. Aksnes 1994. An experimental study on the influence of feeding versus predation risk in the habitat choice of juvenile and adult two-spotted goby *Gobiusculus flavescens* (Fabricius). J. Exp. Mar. Biol. Ecol. 179: 69-79.
Utne, A.C.W., D.L. Aksnes & J. Giske 1993. Food, predation risk and shelter: An experimental study on the distribution of adult *Gobiusculus flavescens*. J. Exp. Mar. Biol. Ecol. 166: 203-216.
Vilhjálmsson, H. 1994. The Icelandic capelin stock. Rit Fiskideildar 13 (1): 1-281.
Werner, E.E. & J.F. Gilliam 1984. The ontogenetic niche and species interactions in size-structured populations. Ann. Rev. Ecol. Syst. 15: 393-425.
Werner, E.E. & Hall, D.J. 1988. The ontogenetic habitat shift in bluegill: The foraging rate-predation risk trade-off. Ecology 69: 1352-1366.

Biological Modelling

By Sigurd Tjelmeland and Bjarte Bogstad.
Institute of Marine Research, P.O. Box 1870 Nordnes, N-5024 Bergen, Norway.

INTRODUCTION

A sound management of fish resources relies upon good answers to the following questions:

1. How much fish is there in the sea today?
2. How much fish will there be in the sea next year and the immediately following years if a given catch quota is taken?
3. How much fish ought there to be in the sea to keep the stock within safe biological limits and - possibly - to utilise the stock optimally?

All these three questions are difficult to answer, basically due to the severe problems connected with measuring the stock size and population structure. The three questions above parallel three distinct steps of management of fish, as the management at present is conducted in the North-east Atlantic:

1. The assessment. Data are collected and the present status of the fish stock in question is evaluated.
2. The short- to medium term prognosis. The fish stock in question is projected forward in time for various options for future catches.
3. The management. Based on option tables from the previous stage a catch quota is decided upon that meets certain criteria for stock development.

For the Barents Sea fish species, the first two steps are performed within ICES while the last step is performed in the mixed Russian-Norwegian fishery commission.

Although there are severe methodical problems and rather large uncertainties connected to the two first steps, the problems are well-defined and quantitatively tractable. As for the last step, well-defined management criteria are lacking both for capelin, herring and cod. The multispecies models developed within the programme «Multispecies Management» shall serve as tools for improving both the short- and medium term prognosis (step 2), as well as the management criteria (step 3).

When the programme started it was also implied that improved management criteria should rest on quantitative knowledge of the interaction between species in the Barents Sea and that the difference in economic value between the species should be taken into account.

THE MODELS MULTSPEC, CAPSEX AND AGGMULT

The main biological multispecies model for the Barents Sea is the MULTSPEC model, where the species included are structured by area, age and length. For some purposes this model is too complex to use, and thus the simpler models CAPSEX (structured only by age and length) and AGGMULT (structured only by age) have been constructed. AGGMULT is also used as an interface to the economic model ECONMULT, developed at Norges Fiskerihøgskole (NFH), Tromsø, which is described elsewhere in this book. In this chapter, we will first present the MULTSPEC model, and then the AGGMULT model, before summarising the use of these models in management and indicating further developments.

Both in forecasting and in management strategy evaluation the economic dimension should be an inherent part of the analysis. In order to analyse the impact of fisheries on different aspects of the economy, the biological models should feed appropriate economic models.

In order to use a model in a bioeconomic analysis there must be a recruitment module. The catch affects the spawning stock and it is only through the effect the catch thereby has on future recruitment we can judge about the «sustainability» of management strategies. Both MULTSPEC and AGGMULT have a recruitment module, but estimation of recruitment parameters has only been done in the AGGMULT model.

CHOICE OF SPECIES

The fish species modelled are cod, herring and capelin. Cod and capelin live in the Barents Sea their whole life span and their whole life cycle is modelled, including recruitment. As for the herring stock, a part of the 0-group drift into the Barents Sea in years of good recruitment and migrates out at the age of 3-4 years. Consequently, no spawning stock-recruitment relation can be modelled using a model that is restricted to the Barents Sea. However, recently MULTSPEC has been connected to a herring model (HERMOD, Dommanes and Hiis Hauge 1994) that covers the Norwegian Sea including Iceland and the Norwegian coast (Bogstad *et al.* 1995a,b).

In addition to the fish species, minke whales and harp seals are modelled in MULTSPEC but not in AGGMULT. However, the data that may be used for parameter estimations are more scant for marine mammals than for fish. The MULTSPEC model has been used to evaluate possible ranges for interactions between marine mammals and fish (Bogstad *et al.* 1995a,b).

In MULTSPEC, plankton is only modelled as a food supply that is variable by area and season, but is constant from year to year. In AGGMULT, 3 different plankton «species» are modelled as separate model entities, preyed upon by fish. These «species» are:

- Small plankton in the southern Barents Sea - food for capelin and herring
- Small plankton in the northern Barents Sea - food for capelin.
- Large plankton organisms - food for cod, herring and capelin.

The choice of species in MULTSPEC and the other multispecies models is based on the hypothesis that total fish production in the Barents Sea area (including Norwegian coastal waters) to a large extent is determined by the development in the stocks of Northeast Arctic cod, Norwegian Spring-Spawning herring and Barents Sea capelin. Accordingly, main emphasis has been placed on modelling the population dynamics of these species, the interactions between them and the parts of the biological and physical environment having a direct and significant impact on their development. The apex predators are an important part of this environment. There are other apex predators than minke whales and harp seals (other marine mammal stocks and birds), and there are other fish stocks which could influence the development of the three modelled species (e.g. polar cod). Another commercially important species which has strong interactions with the cod stock is shrimp, due to predation by cod on shrimp and by-catch of young cod in the shrimp fishery. However, the population dynamics of shrimp is not very well known, and there is presently no TAC on shrimp in the Barents Sea, making inclusion of shrimp in models to be used for management not so urgent. More species may be added to the model later (polar cod is already formally included in the model but has been turned off in all work with the model so far).

Figure 1 shows the calculated consumption by the main predator in the system, cod, of various prey species in the period 1984 to 1995. The basis for the calculations are data from the joint IMR-PINRO stomach content data base (Mehl and Yaragina 1992) and the stomach evacuation rate model developed at NFH, Tromsø (dos Santos and Jobling 1995) within the Multispecies Management programme. Consumption area 1 (see Fig. 2) is not included. More details of the method for consumption calculations can be found in Bogstad and Mehl (1996).

The figure shows a clear inverse relationship between amphipods (large plankton organisms) and capelin as food for cod. As the capelin stock is decimated, as was the case in the period 1986-1989 and since 1993, the amount of capelin in cod stomachs decreases while the amount of amphipods increases. The consumption of herring by cod is relatively low according to the figure, but most of the stomach content data are from fish caught in bottom trawl hauls. The consumption of herring, which mainly is distributed in the upper layers of the sea, is probably underestimated. Cannibalism is important to consider in the assessment and management of the cod stock, as discussed in the 1995 report of the Arctic Fisheries Working Group (Anon. 1996b).

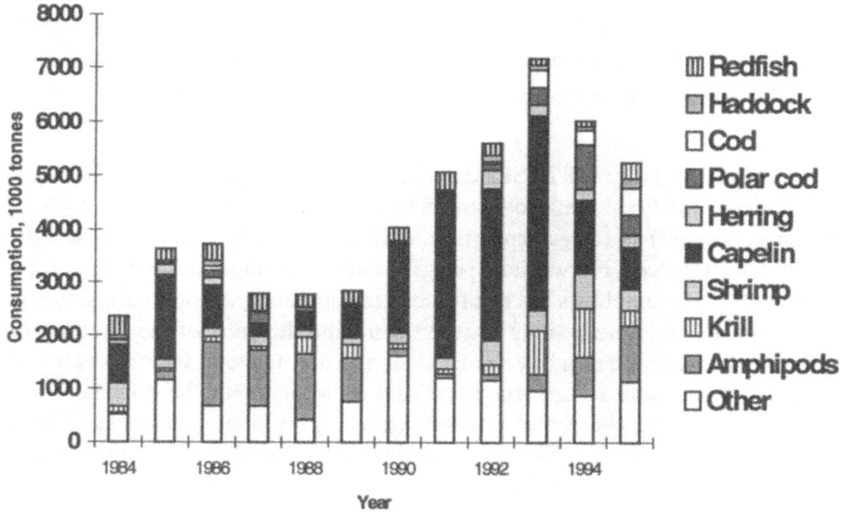

Figure 1. Consumption by cod of different prey species 1984-1995.

THE MULTSPEC MODEL

Main features

MULTSPEC is an area-structured multispecies simulation model for fish and marine mammal stocks in the Barents Sea. It is documented in more detail in Bogstad *et al.* (1995a,b) and Bogstad and Tjelmeland (1995). The species included are cod, capelin, herring, harp seals and minke whales. The stocks which are included in MULTSPEC are represented by tree structures where each level denotes sub-populations distributed over a given dimension, which may be area, sex, age or length. The following division into age groups is used for the various species: capelin 0-5, cod 0-10+, herring 0-15+, harp seals 0-16+, minke whales 0-20+. Capelin is divided into 0.5 cm length groups, herring into 1 cm length groups, and cod into 5 cm length groups. The marine mammal stocks are not divided into length groups. Only capelin and minke whale are divided by sex. The area division used in the model is shown in Figure 2.

As the minke whale stock, depending upon the time of the year, to a large extent is distributed outside the Barents Sea, we have included an 'area 0' in addition to the seven areas in the Barents Sea. The model comprises discontinuous processes like reproduction, which is handled on a yearly basis, and continuous processes like predation and natural mortality, which are performed for each time step. The ordering of the processes is as follows: Migration- Fishing- Predation- Growth. The time step used is one month.

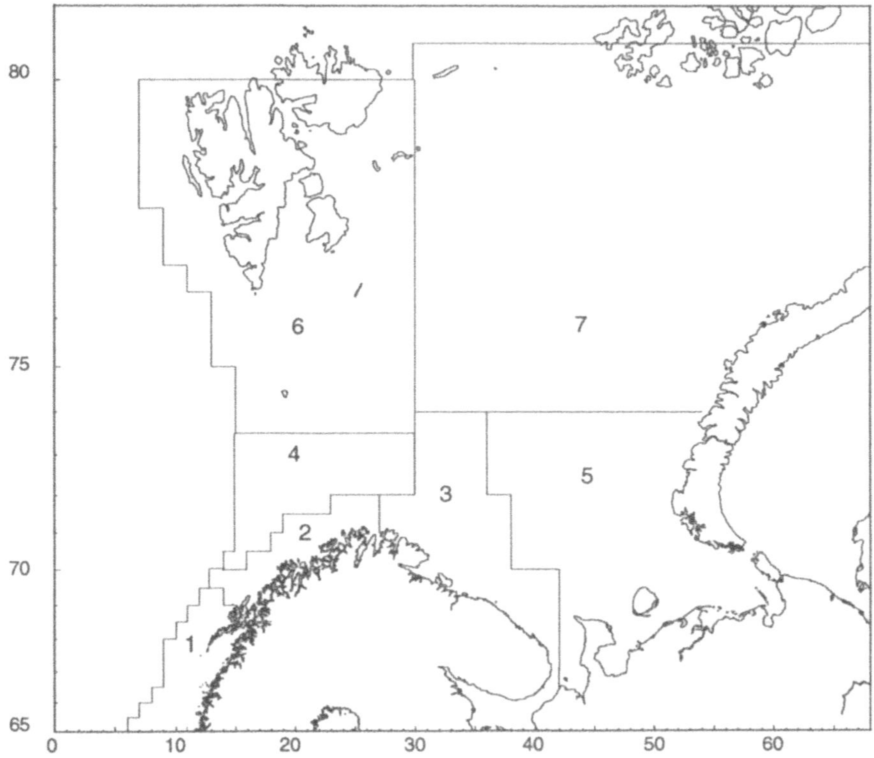

Figure 2. Area division of MULTSPEC.

Stock data

The stock data used in the model is for capelin a joint IMR-PINRO trawl-acoustic survey that is conducted each year in September. For cod direct measurements covering the entire stock do not yet exist, so the number of fish at age from the assessment made by the ICES Arctic Fisheries Working Group is used. The area and length distribution of cod is based on data from the Norwegian winter survey in the Barents Sea. For herring the number of fish at age is taken from the assessment made by the ICES Atlanto-Scandian Herring and Capelin Working group, and the area and length distribution in the Barents Sea is taken from a trawl-acoustic cruise conducted each year in May-June.

The stock abundance at age of harp seals and minke whales are taken from single-species simulations based on various survey and catch data.

Migration

In the Barents Sea the fish and marine mammal species have strong seasonal migrations, and the migration pattern may shift strongly from year to year. The area division shown in Figure 2 is a compromise between the need for many areas in order to describe the most important aspects of migration realistically and the need for few areas for the model to be tractable in terms of computer time and estimation of migration parameters. Also, catch data can only be provided on a relatively coarse area resolution. The present area division suffices to describe the east-west gradient observed in fish growth for all species and the east-west as well as the north-south gradient in migration observed for all species. Migration is modelled using transition matrices constructed based on survey data on area distribution, and is constant over years, i.e. it is not yet a function of oceanographic conditions or food supply.

Maturation

Maturation of fish depends more strongly on length than on age. The number of maturing fish in a length group is defined as a mathematical function of length, increasing from nearly 0 at lengths at which the fish is clearly immature to 1 at lengths at which the fish is clearly mature. The formulation is:

$$m(L) = \frac{1}{1+e^{4P_1(P_2-L)}}$$

where L is the length and P_1 and P_2 parameters.

The maturation function is used to separate the total stock into a mature and an immature part. For capelin this happens October 1, for cod and herring (HERMOD) January 1. In MULTSPEC the function is at present used for capelin only. For capelin the parameter values have been estimated using the length distributed model CAPSEX (Tjelmeland and Bogstad 1993). For cod all fish present in area 1 at April 1 is assumed to be mature, and the migration parameters are set so that the proportion of the stock found in this area at this date is approximately the same as the proportion mature calculated from the maturity ogives used in the assessment. For marine mammals a fixed proportion of mature individuals by age is used.

Catch

For capelin, historical data on catch by area, month, age and length have been prepared for use in the model. For herring and cod, such data are in preparation. For marine mammals, data on catch in numbers for each year are available. For runs into the future, a fixed fishing pattern by age (cod), by length (herring) or fishing mortality variable between mature and immature stock (capelin) is used. For marine mammals, the catch is a fixed percentage of the stock. For harp seals, the catch rate applied to 0-year-olds is different from the catch rate on older age groups.

Recruitment

For cod and capelin a Beverton-Holt spawning stock-recruitment relationship is used:

$$\text{Recruitment} = P_{13} \frac{SSB}{P_{14} + SSB}$$

where SSB denotes the spawning stock biomass.

For herring, a depensatory stock-recruitment relationship is used in HERMOD:

$$\mathrm{Re}\,cruitment = H_3(1 - e^{(e^{-0.17377} - e^{(0.00005(SSB-2500))})})$$

The spawning stock is measured in thousand tonnes, so this implies that the spawning stock-recruitment relationship has an inflection point at 2.5 million tonnes. Two values of H_3 are used, corresponding to good and bad recruitment years.

For marine mammals, the fertility is assumed to be constant. MULTSPEC has an option for allowing the fertility to be dependent on the food supply for marine mammals, but this option has not been used in any of the simulations presented here.

Predation

The predation is modelled in the same way for fish that eat fish (cod) and plankton feeders (herring and capelin). In MULTSPEC, the plankton is modelled as an exogenous food abundance field that varies with area and time of the year.

First the area density of food available to the predator - φ - is calculated. The availability for a prey item as source of food for the predator is found by multiplying the prey abundance with a scaling function, called the suitability of the prey as food for the predator. The suitability has a species dependent term that incorporates the effect of processes like, for instance, schooling behaviour of the prey and a size dependent term that describes that a small predator cannot eat a large prey item. In addition to the food supply from modelled species a parameter for other food for cod and marine mammals is used.

$$\varphi = \sum_{speciessize} Suitability_{species} \times Suitability_{size} \times \frac{\mathrm{Pr}\,ey}{Area} + P_{23}$$

P_{23} is the concentration of non-modelled food, estimated from the stomach content data and is determined by the ratio of non-modelled to modelled food in the stomachs. The summation is made over all prey species groups (prey species and size groups). Prey is prey biomass within each prey species group.

The size-specific suitability is determined by linear interpolation in a length by length parameter matrix for each prey species - predator species combination. These matrices are determined by informed guesses based on information from the stomach content data base.

The next step is to calculate the feeding level, defined as:

$$\mathrm{f}(\varphi) = \frac{\varphi}{\mathrm{P}_3 + \varphi}$$

At low values of the food density, this function is proportional to the food density. For high values of the food density, the function approaches the limiting value 1, at which there is feeding to satiation. The parameter P_3 is the food density at which the predator feeds at half the maximum rate.

The feeding level is multiplied with the maximum consumption for each predator and with the number of predators. This gives the total consumption which is distributed on the prey species according to the proportion in the diet.

The parameter P_3 for cod is estimated in MULTSPEC for the situation of cod preying on mature capelin during the spawning period January-March.

The total consumption by n(a) predators of age a (with individual weight wla) is given as:

$$\text{Consumption} = P_{22}(1) \times f(\varphi) \times e^{0.104 \times T - 0.000112 \times T^3} \times w(a)^{P_{22}(2)} \times n(a)$$

where T is the temperature (°C). $P_{22}(1)$ for cod is estimated by MULTSPEC, $P_{22}(2)$ and the temperature dependency from the literature (Jobling 1988).

Bioenergetic research might give a quantitative relation between the parameters P_3 and P_{22} (food intake) on the one hand and the parameters P_4, P_5, P_8, P_9 and P_{10} (growth, see below) on the other hand. This possibility has not yet been investigated.

For marine mammals, the energy consumption is assumed to be a function of the body weight and variable by season, but independent of the food density.

Natural mortality

In addition to the mortality generated by consumption from predators and catch, the model allows for a residual natural mortality, which may be dependent on age. In MULTSPEC, 0-3 year old herring generates a natural mortality on 0-group capelin.

Growth

For the fish species, the growth in length is modelled in MULTSPEC as:

$$\text{GrowthLength} = P_4 \times f(\varphi) \times L^{P_5}$$

and the growth in weight is modelled as:

$$\text{GrowthWeight} = P_8 \times (f(\varphi) - P_{10}) \times L^{P_9}$$

The growth of marine mammals can be set constant, or it can be made dependent on the food supply. In the simulations carried out in Bogstad *et al.* (1995b), constant growth was used.

Parameter estimations

Tjelmeland and Bogstad (1993) estimated the maturation parameters for capelin using the CAPSEX model by comparing the age distributions of capelin from two consecutive autumn acoustic surveys. Bogstad and Tjelmeland (1995) estimated the parameters P_3 and $P_{22}(1)$ for the cod-mature capelin subsystem, using the maturation parameters for capelin estimated by Tjelmeland and Bogstad (1993). The modelled consumption of mature capelin by cod was compared to the consumption calculated directly from the stomach content data using the stomach evacuation rate model developed by dos Santos and Jobling (1995). This was done for each of the MULTSPEC areas 2-5 in the months January-March, and a maximum likelihood function was used to find the best fit. It was found that an annual migration pattern of capelin also had to be estimated in order to obtain a good fit.

These MULTSPEC results as well as the MULTSPEC model itself have been reviewed by the ICES Multispecies Assessment Working Group (MSA) (Anon. 1996a) along with the structural similar model for Icelandic waters, BORMICON, and the multispecies VPA model. The MSA recommended that MULTSPEC be used in parameterising the model that shall be used for calculating capelin consumption by cod during the management process for capelin.

During the 1995 meeting of the MSA also initial comparisons of modelled to observed cod growth were made. This analysis will in the future be extended to a formal estimation. To the extent possible, further work with MULTSPEC will be done through the ICES co-operation.

Simulation results

The results of the baseline simulation (reference run) carried out in Bogstad *et al.* (1995b), are shown in Figure 3. The figure shows only the development of the cod, capelin and herring stocks. The catch of marine mammals was set so that these stocks stabilised at values close to the starting level (80 000 minke whales and 600 000 harp seals) . The simulations were carried out for the period 1992-2012. For herring, a recruitment pattern of 2 strong year classes followed by 6 weak ones was assumed. Since the 1991 and 1992 year classes of herring are strong, it was assumed that the next 6 year classes would be weak. Thus, the 1999, 2000, 2007 and 2008 year classes of herring were assumed to be strong. The ratio between the strength of strong and weak year classes (as 0-group on July 1) was set to 12.5. The figure shows that the strong year classes of herring causes capelin to decline sharply, which again infers reduction of the cod biomass due to reduced growth. The figure also shows that the capelin biomass increases rapidly when the herring biomass in the Barents Sea (B.Her) decreases to a low level.

Several other simulations were carried out, studying the effect of 1: removing the marine mammal stocks, 2: no catch of marine mammals 3: varying the food preferences of marine mammals and 4: varying the cod stock's preference for herring as food.

The main effects may be summarised as follows:

The herring stock increases or decreases as predation from marine mammals decreases or increases. With suitability's as in the reference run, the herring stock is much more sensitive to changes in the minke whale stock than to changes in the harp seal stock. The development in the capelin stock is mainly determined by changes in the herring and cod stocks. The effect on capelin of changes in these stocks generally goes in the opposite direction of effects from changes in marine mammals predation on capelin. This results in an increase or decrease in the capelin stock when the minke whale stock increases or decrease i.e. a counterintuitive effect. Since herring is less sensitive to changes in the seal stock than to changes in the minke whale stock, and since predation from seals on capelin is high, an increase or decrease in the seal stock leads to a decrease or increase in the capelin stock, i.e. the effect intuitively expected.

The effects on the cod stock from changes in the marine mammal stocks are difficult to summarise. Generally, the cod stock will increase or decrease when the mammal

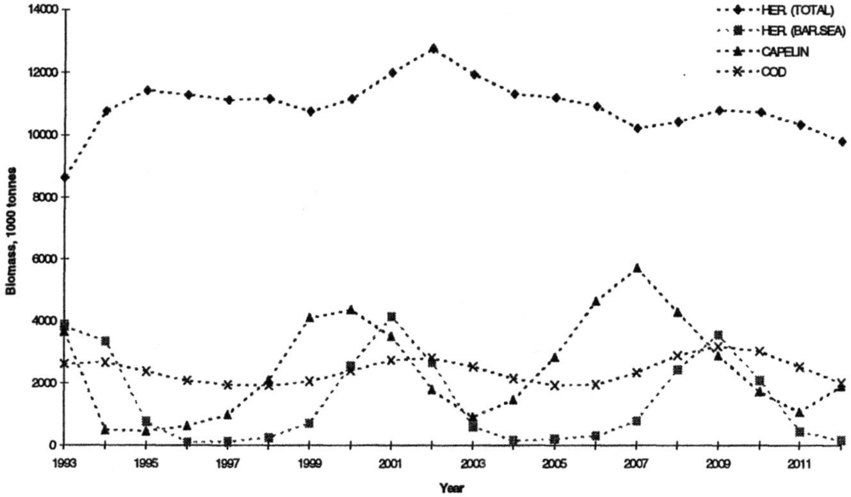

Figure 3. Stock development in the MULTSPEC reference run.

stocks decrease or increase, as intuitively expected. For example, if the seal population is not exploited and is allowed to increase, mean annual cod catch in the last year will be 112 000 tonnes lower than in the reference run. This would be a substantial loss to the fishery taking into account the high value of cod. However, because of the strong cod-capelin interactions, resulting in a tendency of cyclic variations in the two stocks with a time lag between the two stock trajectories, the changes in the cod stock may in some years be in the opposite direction than expected when compared to the reference run.

One interesting feature which again reflects the complexity of the system is that there would be larger gains on average in the cod fishery by removing the seals than by removing the whales, despite the fact that whales eat more cod than seals do in the reference run. The explanation lies in the herring-capelin-cod dynamics: Removing whales have a large effect on the herring stock, leading to strongly reduced capelin stock and thereby reduced cod growth.

Decreasing the suitability of herring as food for cod had much larger effects than changing some of the marine mammals preferences and even more dramatic effects than removing both marine mammal stocks from the system. The herring stock increased above historic levels, with resulting detrimental effects on the capelin stock. Also the cod stock decreased due to low capelin stock.

THE AGGMULT MODEL

MULTSPEC is because of the area distribution too complex and computer time consuming to at present be a useful tool for the repeated stochastic stock simulations that are needed in order to evaluate multispecies management rules under uncertainty. Therefore, as stated in the beginning of section 2, a simplified version - the AGGMULT model - was constructed to serve as a bridge between the biological modelling in the MULTSPEC project and bioeconomic modelling at NFH.

AGGMULT is documented in detail in Tjelmeland (1996). The losses in moving from the area distributed model MULTSPEC to the area integrated model AGGMULT are twofold:

1 - The predation is dependent on the overlap between the species and the consumption of one species on the other, where the latter depends on the concentration of the prey species. In an area disaggregated model these two fundamentally different processes may be kept separate from each other, enabling a simpler and more intuitively appealing modelling of the predation.
2 - The parameter P_3 in the feeding level function is difficult to compare between the models, since in the area aggregated model this parameter will depend on the formulation of the concentration as function of total biomass.

Division in time and stock structure

In AGGMULT there is only division on age, and the model uses one quarter (three month) time step. The same age groups as in MULTSPEC are used.

Maturation

The same maturation function as in MULTSPEC is used. However, AGGMULT does not have a length dimension. For cod the mean length by age is used and the maturation parameters are estimated using empirical maturation data from The Arctic Fisheries Working Group. These data are in the form of percentage maturation by age, so-called maturity ogives.

For capelin the real historical length distribution is read in and the parameter values previously estimated using the length distributed model CAPSEX (Tjelmeland and Bogstad 1993) are used. The fraction mature within each age group is calculated for each sex separately, and the mean maturation fraction is then applied in AGGMULT. For runs into the future the length distributions for the scenario year are used, see the section about stochastic simulation.

Predation

The predation is modelled in the same way in AGGMULT as in MULTSPEC, but plankton is modelled dynamically in AGGMULT, in contrast to MULTSPEC. The plankton biomass model used is described in the growth section. The dimension of P_3 is 1000 tonnes pr square nautical mile. In order to convert the prey stocks from absolute abundance (million tonnes) to concentration in AGGMULT, the Barents Sea is assumed to have an effective size of 62 000 square nautical miles.

The parameter P_3 is estimated in MULTSPEC for the situation of cod preying on mature capelin during the spawning period January-March. In this period the capelin appears in large schools. There is reason to believe that P_3 might be different in the summer-autumn situation where the capelin is distributed over wide areas in layers or smaller schools. The problem is basically whether a smaller fraction of the cod stock preying on high concentration or a larger fraction of the cod stock preying on smaller concentration might have the same effect on the capelin stock. The value of P_3 to use is dependent on how the overlap between the species is defined. This point is crucial in both MULTSPEC and AGGMULT and is also a key issue when comparing results from the area disaggregated model MULTSPEC to those of the area aggregated model AGGMULT.

In AGGMULT, the same value of P_3 as estimated by MULTSPEC (0.009) is used for cod preying on capelin during January-March, while we have experimented with two different values of P_3 (denoted by FEED) the rest of the year, FEED = 0.009 and FEED = 0.05. Somewhat arbitrarily, the P_3 value for plankton feeders is set to 2.0*FEED.

The size suitability is an age by age parameter matrix for each prey-predator combination and is not estimated from data, rather determined by informed guesses. The relative species suitability is estimated from stomach content data, both for cod and plankton feeders.

The total consumption for cod is given in the same way as in MULTSPEC. In AGGMULT, $P_{22}(1)$ for plankton feeders is estimated with the model using stomach content data and an evacuation rate of 0.1 gram per hour regardless of body weight. This value is higher than values previously calculated for capelin (Ajiad and Puschaeva 1992) and presumably lower than the evacuation rate for herring. The model used is very coarse and in later versions of AGGMULT care will be taken to utilise experiments of evacuation rates for herring, as well as attempting to make the model realistically body size dependent. Before good empirical data are available, sensitivity analysis should be made trying different models.

Growth

In AGGMULT the growth equations are simplified:

$$\text{GrowthLength} = f(\varphi) \times P_4(\text{species, age})$$

$$\text{GrowthWeight} = f(\varphi) \times P_8(\text{species, age})$$

P_4 has the dimension cm pr quarter and P_8 kg pr quarter. f is the feeding level defined above.

For plankton the following growth equation is applied:

$$\text{IncreaseBiomass} = \text{plprod} \times (\text{PLANK} \times \text{pl max} - \text{plbio})$$

where plbio is the biomass of the plankton species and plprod and plmax are quarter-dependent but yearly constant parameters for each plankton species. PLANK is a scaling variable introduced to test the model for different levels of plankton abundance.

Recruitment

Recruits enter into the simulation at the beginning of the third quarter of the calendar year.

We assume that the maximum value of the recruitment (P_{13}) is somewhat higher than the highest estimated value, i.e.

$$P_{13} = \text{REC} \times \text{Max(Estimated recruitment)}$$

The scaling factor REC is unknown. Runs are being made with different values of this parameter. P_{14} is calculated each year by using the estimated recruitment in the scenario year.

The idea behind this formulation is that there may well be a strong relationship between recruitment and spawning stock *in one and the same year* in the sense that if the spawning stock one particular year had been larger the recruitment would also have been larger. The large observed variability stems from strong environmental influence that makes this relationship different from year to year. One relationship valid for all years may be difficult or impossible to find.

The basic idea behind placing the uncertainty on the maximum recruitment is that the physical-biological processes behind the recruitment as function of spawners may be different at small levels of the spawning biomass from the processes determining the limiting value of recruits at large values of the spawning biomass. The strongest environmental influence is thought to act on the proportional part of the recruitment curve, rather than on the limiting part at high spawning stock values. Thus, the impact of the varying environmental conditions are calculated year by year without making any further assumptions.

Of course, in a later version of the model the opposite should be tried and results compared.

In AGGMULT the environmental variability is modelled through variations in P_{14}. It is a research task to find a relation between the yearly calculated P_{14} and environmental conditions.

Migration

In the present implementation of AGGMULT migration is simply disregarded, although some of the knowledge about the geographical distribution of the species is taken into account, e.g. by not letting cod prey on mature herring. The plans for the next AGGMULT version include modelling an overlap index based on empirical studies of overlap. Also, the parameter P_3 will be modelled more realistically in the next version of the model.

Parameter estimation

During the estimation of the model a level of recruitment is found each year that when propagated through the model accounts for the observed stock size at the recruitment age, which is 3 years for cod, one year for herring and 2 years for capelin.

Cod cannibalism that occurs before 3 years automatically is dealt with in the model. However, cannibalism on cod aged 3 years and older - which occur in some years (Anon. 1996b) - is difficult to deal with because the $M = 0.2$ assumption used in the ICES VPA procedure that generates the stock data used in the model is violated.

Both the growth parameters P_4 and P_8 are in AGGMULT estimated year by year. This is also the case with the residual natural mortality.

The stomach content data are used to estimate the parameters in the model that govern the predation. Also stomach content data for plankton feeders are used. For predation by herring on capelin the amount of data is scarce. Some support for the model is taken from an empirical study (Huse and Toresen 1995). In AGGMULT it is assumed that the suitability of capelin as food is the same for 0-group herring as for older herring and that the predation is on 0-group capelin only.

When FEED and PLANK are changed, all the other parameters have to be re-estimated.

Stochastic simulation

The main simulation mode is stochastic. During runs into the future the yearly varying parameters are drawn at random from the time series of historically estimated values. In the present report this is done in the simplest possible way: Each year a year in the historical time series is drawn and the parameter values estimated for that year are used. This procedure covers all the uncertainty connected with the estimation of parameters in AGGMULT, since nearly all the parameters are estimated on a year to year basis. However, there might be reason to believe there exists a substantial autocorrelation, i.e. the growth parameters one year may not be totally independent of the growth parameter the previous year due to autocorrelation in factors affecting growth (temperature). Therefore the uncertainty applied in the model might be too large. Also, one might expect some intra-annual correlation between parameters, which is not captured in the procedure applied. It is a project for the future to look closely into possible relationships between the yearly estimated parameters and environmental indices. Such relationships might reduce uncertainty and help building submodels for processes that at present are modelled just as yearly estimated parameters.

The stochastic procedure applied does not capture uncertainty related to parameters that are constant over years. The effect of uncertainty in these parameters must be investigated by reparameterising the model for several different parameter values and evaluate consequences in each case. In the present report we have concentrated on the processes that presumably are the most uncertain ones, connected to the parameters FEED, PLANK and REC.

Table 1 shows the mean biomass for runs over 50 years using different values of FEED, PLANK and REC. As a test of the model's extreme situation behaviour, in addition to runs with all three interacting species, runs with only two and runs with only one species have been performed. In those cases, the other species are deleted from the model entirely.

For cod and capelin, the total biomass is shown, while for herring, only the immature biomass is shown. During all these runs one and the same sequence of stochastic events has been used and there has been no fishing. The parameters have been drawn from the historic period 1973-1991. The reason for stopping in 1991 is that the recruitment of cod could not be estimated for later years, since the recruitment age is 3 years and the last year of data used was 1994. The recruitment to the herring stock has been restricted backwards in time to 1983, since we believe that the period prior to 1983 was abnormal and may well never return. The recruitment conditions of herring were not drawn randomly, rather the time series was used cyclically, thus assuming no correlation between recruitment of herring and other parameters in the model.

Different assumptions concerning herring recruitment in relation to other parameters in the model will be explored during the implementation of the model (see below).

During the parameter estimation there has been no concern for the model's long-term behaviour, i.e. no parameters have been «tuned» to yield acceptable long-term behaviour. Therefore, running the model into the future and observing its characteristics provides an independent test of the model.

For all combinations of the unknown parameters and for all combinations of stocks the model behaves well in the sense that the stocks are inside reasonable limits. A model that allows for 30 million tonnes of cod or a model that leads one or more species to extinction certainly will have some weak spots. The results given above give confidence in further work with the model.

It is seen that the stocks indeed interact with one another. The cod controls capelin stronger than does herring. An increase in FEED affects capelin through its effect on cod. In presence of cod an increase in FEED leads to an increase in capelin because the cod stock is decreased. In absence of cod an increase in FEED leads to a decrease of capelin, which is what one would expect because an increased FEED means lower grazing capability for all species.

The uncertainty in the knowledge about the scaling of plankton abundance quantified in the PLANK variable is of minor importance. The reason for this is that changes in PLANK lead to changes in estimated parameters (growth, suitability) in order to fit the fixed length and weight and stomach content data used in the estimation process.

An increase of the REC parameter from 1.1 to 1.5 leads to moderate increases of cod and capelin due to stronger recruitment. Increasing REC further to 5.0 leads to a substantial increase in the cod stock. If there is only capelin a REC-value of 5.0 triples

Table 1. Simulation results (mean biomass, million tonnes) for cod, capelin and herring (indicated in the second line in the heading) when different species combinations (first line in table heading) are included.

			All	All	All	CodCap	CodCap	HerCap	HerCap	Cod	Cap
FEED	PLANK	REC	Cod	Her	Cap	Cod	Cap	Her	Cap	Cod	Cap
0.009	0.5	1.1	2.63	0.19	3.26	2.66	4.24	0.26	6.02	2.46	7.86
0.009	0.5	1.5	2.95	0.18	3.23	3.00	4.34	0.26	7.56	2.76	9.72
0.009	0.5	5.0	4.41	0.16	3.20	4.49	4.25	0.26	20.95	4.13	24.42
0.009	1.0	1.1	2.56	0.18	3.21	2.58	3.97	0.26	5.98	2.39	7.41
0.009	1.0	1.5	2.87	0.18	3.14	2.91	4.02	0.26	7.44	2.68	9.17
0.009	1.0	5.0	4.27	0.16	3.03	4.34	3.91	0.26	20.50	4.00	23.52
0.05	0.5	1.1	2.42	0.20	3.69	2.47	4.95	0.27	5.41	2.24	7.61
0.05	0.5	1.5	2.76	0.19	3.91	2.83	5.31	0.27	6.65	2.54	9.22
0.05	0.5	5.0	4.31	0.17	4.18	4.42	5.69	0.27	15.47	3.94	19.65
0.05	1.0	1.1	2.24	2.48	1.07	2.27	5.19	3.86	1.09	2.05	7.86
0.05	1.0	1.5	2.54	2.34	1.05	2.60	5.68	3.87	1.08	2.33	9.71
0.05	1.0	5.0	4.00	1.83	1.46	4.20	6.96	3.89	1.77	3.66	23.12

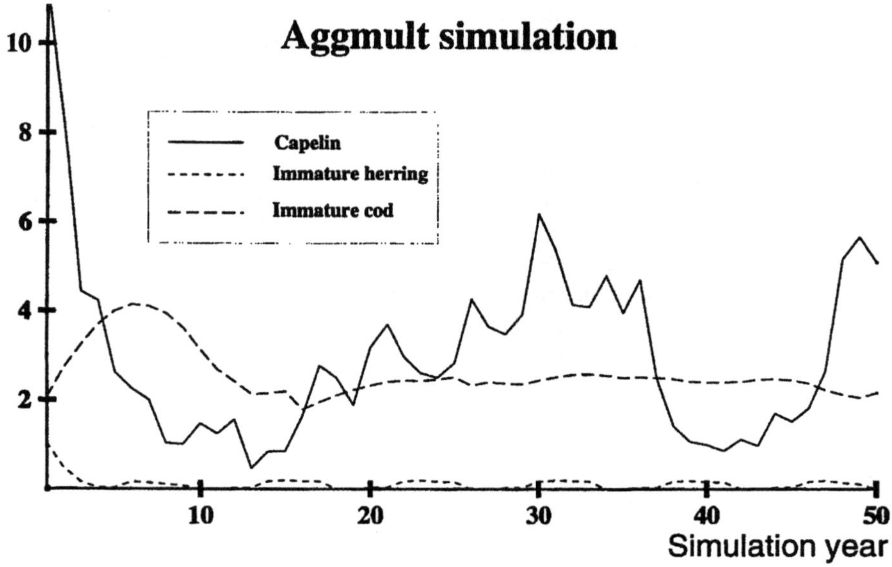

Figure 4. Cod, herring and capelin for one stochastic sequence. Y-axis in mill. tonnes.

the capelin stock with respect to a REC-value of 1.1. If cod is present the increase is slight or even negative, due to the controlling ability of the cod.

Figure 4 shows the biomass of cod, herring and capelin for the first 50 years for one stochastic sequence where FEED = 0.009, PLANK = 1.0 and REC = 1.1.

The model will be analysed further with a richer suite of model and parameter uncertainties, including:

- Modifications of the use of the Beverton-Holt recruitment function.
- The relation of P_3 for plankton feeders to P_3 for cod.
- Modelling of the overlap between the species.
- The formulation of the feeding level function.
- Size-dependent suitability.
- Intra-annual variation of plankton parameters.
- Regressing yearly estimated parameters to environmental conditions.
- A more elaborate evacuation rate model for plankton feeders.

The stochastic nature of the model is revealed when repeated runs are made each time using a different stochastic sequence. Figures 5-7 show the biomass of cod, herring and capelin for 20 different stochastic sequences using the same assumptions as earlier. All runs are made with no catch and for FEED = 0.009, PLANK = 1.0 and REC = 1.1.

Although varying, the cod stock biomass is well confined between 2.0 and 2.5 million tonnes after ca. 10 years, despite the substantial uncertainty in the recruitment.

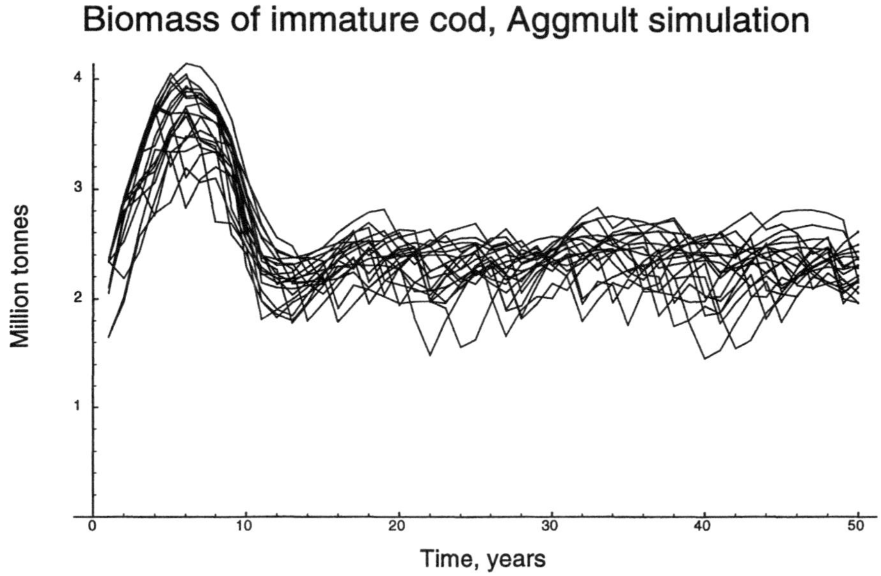

Figure 5. Stochastic simulation of cod.

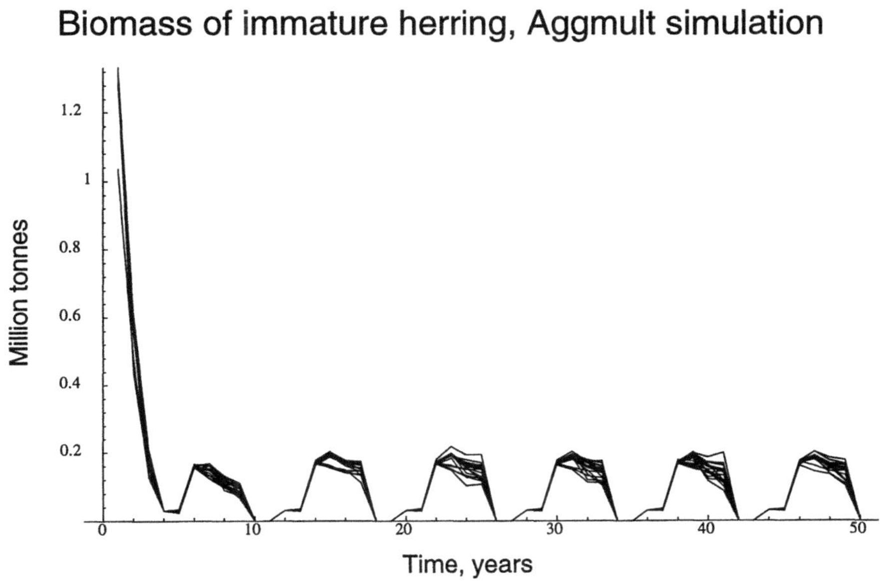

Figure 6. Stochastic simulation of herring.

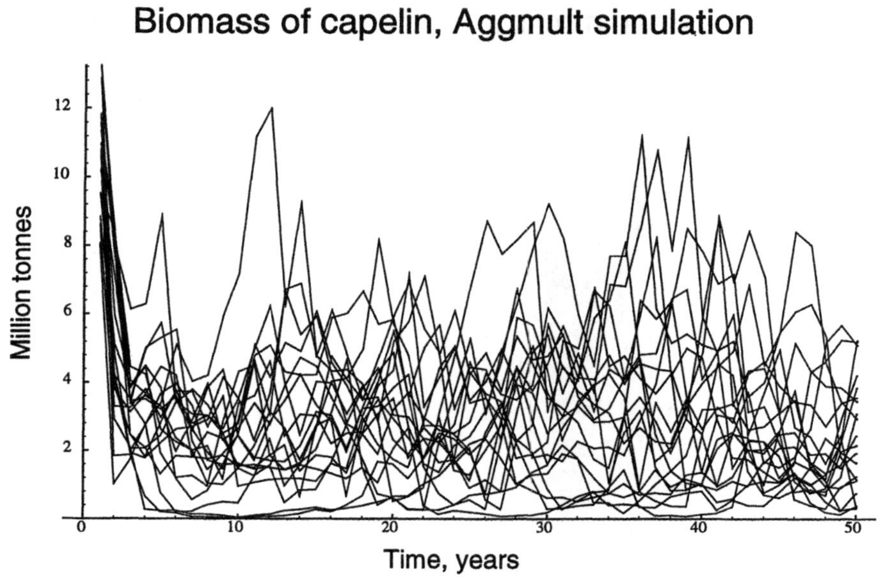

Figure 7. Stochastic simulation of capelin.

Thus, the model does not seem to reproduce the variability observed.

The recruitment conditions for herring are not drawn randomly, yet the time series exhibits some stochastic fluctuations. This stems from variable growth and variable predation on herring from cod. The herring abundance is much lower than one might expect.

The capelin stock fluctuates wildly, and it is difficult to see the signals from the cyclic herring stock, indicating that the predation from herring on capelin is somewhat too small in the model. This could stem from the very coarse assumptions about the evacuation rate.

Using AGGMULT to study management strategies

The assessment of fish stocks today is based on well-defined and controllable routines. However, management in the sense of selection among the forecast options from the assessment is not well-defined. Even if setting a quota is a highly political process involving different economic and sociological values in each country, one should expect that once these options were specified, the quota should be given in a straight-forward and controllable process. For instance, if the objective for capelin management was to maximise long-term yield in biomass terms, the quota should be given once the data were updated each year. The role of multispecies modelling is to specify the quota that best would meet the different objectives for different species given by management bodies.

The New York convention on straddling stocks emphasises the precautionary principle in the management of internationally shared fish stocks (Anon. 1995). As there is now a growing tendency for the ICES assessment working groups to give risk-related medium term forecasts, application of the precautionary principle implies that the larger the uncertainty, the smaller the quota.

The medium-term forecasts relate to some well-defined management rule, either a target biological reference point (TBRP) or a limit biological reference point (LBRP). These might, for instance, be a fixed fishing mortality or a fixed quota. In (Anon. 1994) a reference point is defined as:

> "A reference point is considered to be a conventional value derived from technical analysis which represent a state of fishery and population, and is intended as a guide for fisheries management".

The requirement that a reference point shall be derived from "technical analysis" implies that a reference point must be referred to a particular population dynamic model of the fish stock(s) in question.

The main application of the Aggmult model is to shed some light on what TBRPs and LBRPs might be in a multispecies management of the Barents Sea. The uncertainty connected to the estimated processes will not lead to uncertainty in the estimated TBRP, the length of the simulation time or number of replicate simulation series could always be made large enough to estimate the TBRP with any required precision. This is, however, not the case with the uncertainty connected to the model itself or to parameters not estimated from data. In the present version of the Aggmult model, such parameters are PLANK, FEED and REC. The precautionary principle could then be applied by estimating TBRPs for a wide range of these parameters and selecting the most conservative, a principle that we will coin the model-uncertainty precautionary principle.

Management rule

The first step is to suggest a management rule, i.e. the quota as a function of the data. If the goal is to maximise the yield from the cod stock such a rule might be to use a fixed fishing mortality for cod and no fishing on capelin. One could also use a constant spawning stock or more elaborate and adaptive criteria.

Success rule (objective function)

In addition one must formulate a measure for the success of management. If the goal is to maximise the yield from the cod stock this measure might simply be the long-term mean catch. In addition one might use safeguarding measures as a low probability of the stock being small over extended periods of time. The latter measure could be applied as a constraint or be weighted against the former.

Simulation, estimation of the management rule

When a management rule containing one or more unknown parameters and a success rule have been formulated the model is run over several years where in each year the management rule is applied and the success rule accumulated. This is repeated many times for different values of the parameter(s) in the management rule. At the end one finds the parameter(s) that maximises the success measure.

It is essential that the stochastic uncertainty is rooted in data. If means and standard deviations in distribution functions are thrown in from more or less qualified guesses the estimated management rule would be subject to a supposedly large and uncontrollable uncertainty that would render it - in the worst case - useless.

Model uncertainty

The model formulation itself might be uncertain on several points. In the AGGMULT case this uncertainty is most of all linked to the parameters FEED, REC and PLANK. Uncertainty connected to FEED and PLANK must be handled differently from the uncertainty connected to the estimated parameters because the estimation results are dependent on the value of these parameters. For reasons of computer time it is not possible to investigate many combinations of these parameters.

Example of estimation of management rules using AGGMULT

The method outlined above has previously been applied using the model CAPSEX (Tjelmeland and Bogstad 1993) which has been made a multispecies model by including a dynamic cod compartment and herring influence on capelin recruitment through the Beverton-Holt half value. However, the success measure applied was far more complex than any of those suggested above.

As an illustration of the use of AGGMULT the method described above is applied for cod using AGGMULT as simulation model. The runs are over 75 years and the success measure is accumulated for the last 45 years to avoid the effect of initial transients. The same weights at age for the catch as for the stock has been used, which may be somewhat unrealistic.

The management rule is a constant fishing mortality and the success measure is maximum yield. In other words, F_{MSY} (that in (Anon 1994) is treated as an example of a LBRP) is sought. Table 2 summarises the results.

With the exception of REC = 5.0 the estimated F-values are more or less of the same magnitude as the F-values applied during management of the cod stock in recent

Table 2. Estimated optimal fishing mortalities ($F_{5\text{-}10}$) and associated mean catch (million tonnes) for cod using AGGMULT.

FEED	PLANK	REC	Fopt	Catch
0.009	0.5	1.1	0.508	0.268
0.009	0.5	1.5	0.573	0.377
0.009	0.5	5.0	0.960	1.223
0.009	1.0	1.1	0.486	0.254
0.009	1.0	1.5	0.554	0.357
0.009	1.0	5.0	0.960	1.168
0.050	0.5	1.1	0.425	0.216
0.050	0.5	1.5	0.513	0.313
0.050	0.5	5.0	0.960	1.053
0.050	1.0	1.1	0.364	0.189
0.050	1.0	1.5	0.471	0.280
0.050	1.0	5.0	0.960	0.980

years. The estimated F-values increase with increasing value of REC indicating that for high stock values the effects of cannibalism outweighs the effect of the spawning stock. There is a variation with FEED, but not dramatic. The change in the estimated F-value with PLANK is rather small.

The associated mean catch is smaller than one would expect, partly or wholly because equal individual weights in catch and stock are used. There is a correlation between high estimated optimal F-value and high mean catch.

The above findings combined with the model-uncertainty precautionary rule indicate an optimal fishing mortality of Barents Sea cod of about 0.40, somewhat lower than the present fishing level. This analysis should be extended to include the effect of using a fishing mortality derived for one set of the uncertainty parameters FEED, PLANK and REC on simulations using another set of these parameters. The full analysis is complicated and time consuming and will be deferred to later upgradings of the AGGMULT model.

This exercise is meant as an illustration only. As has been shown there are large uncertainties connected to imperfect submodels. However, it is possible to convert the basic uncertainties in the various processes and data sources to uncertainties in the final product, i.e. the quota that may be advised in a given situation.

In the AGGMULT model, effects of results from ongoing biological research on management decisions can be tested. Thus the model may serve as a bridge between managers, biologists and modellers.

CONCLUSIONS

In this chapter the biological models developed within the programme 'Multispecies management' have been described, and the use of the models so far has been presented.

Although results from the models described here have not yet been implemented directly in management, results from the research programme have definitely helped to improve the medium-term prognosis of cod, capelin and herring abundance given in the last years through a better understanding of the interactions between cod, capelin and herring. For cod, the prognosis of individual growth made in the last years have been based on knowledge about expected capelin abundance and on bioenergetic studies. Also, cod cannibalism, the magnitude of which has been calculated from the stomach content data, has been taken into account. For capelin, the recruitment has been predicted to be bad as long as there is much herring in the Barents Sea. Also, an upper bound on the consumption of mature capelin by cod has been calculated based on stomach content data (Bogstad and Gjøsæter 1994). The prognosis made for cod growth and capelin recruitment based on this 'qualitative' knowledge have so far proved to be fairly good. Cod cannibalism was introduced in the assessment and the prognosis for the first time in 1995, so the predictive value is still too early to evaluate. However, it was encouraging to note that including cannibalism gave a better fit between the abundance indices from surveys and the VPA for age groups 1-3.

There is still much work to be done in quantifying the cod-herring-capelin interactions, and the work on such interactions must still have the highest priority. At the same time, one should seek to improve the links to oceanography, plankton, marine mammals and economy. To do this at the same time will require considerable effort.

From 1996 onwards, the consumption of mature capelin by cod will be calculated by the ICES Working Group using the predation equation given in this paper with parameter values based on MULTSPEC estimations. We also expect that the mortality of immature capelin and the growth of cod soon will be predicted using results from MULTSPEC and AGGMULT directly.

All ICES Assessment Working Groups are now asked to perform sensitivity analysis and to provide advice on management objectives, as well as taking multispecies interactions into account, and the modelling work presented in this chapter will obviously be of relevance in this context. It is important for the credibility of the management that the uncertainty estimates are rooted in the data. Speculative uncertainties will discredit the management strategies obtained as well as the quota decisions that are made on basis of the model.

As for developing management criteria, the example given for cod in this chapter indicates how one may proceed towards obtaining a multispecies management strategy. In the development of such a strategy, one may also make use of the work done in the 'Scenario Barents Sea' project described elsewhere in this book.

Furthermore, for sound management good biological models alone do not suffice, even if coupled with adequate economic models. The models should also be understood and usable by the decision makers. This is a difficult task to accomplish even for simple models, let alone complex multispecies models like the MULTSPEC or even more modest models like the AGGMULT model.

REFERENCES

Ajiad, A.M. & T.Ya. Pushchaeva 1992. The daily feeding dynamics in various length groups of the Barents Sea capelin. Pp. 181-192 in B. Bogstad and S. Tjelmeland (ed): Interrelations between fish populations in the Barents Sea. Proc. fifth PINRO-IMR Symp. Murmansk, 12-16 August 1991. Institute of Marine Research, Bergen, Norway.

Anon. 1996a. Report of the Multispecies Assessment Working group. ICES C.M. 1996/Assess:3, 183 pp.

Anon. 1996b. Report of the Arctic Fisheries Working Group. ICES C.M. 1996/Assess:4, 311pp.

Anon. 1995. Draft agreement for the implementation of the provisions of the United Nations convention on the law of the sea of 10 december 1982 relating to the conservation and management of straddling fish stocks and highly migratory fish stocks. A/CONF.164/33. United Nations conference on straddling fish stocks and highly migratory fish stocks. New York, 3 August 1995.

Anon. 1994. Reference points for fisheries management: Their potential application to straddling and highly migratory resources. United Nations conference on straddling and highly migratory fish stocks. A/CONF.164/INF/9. New York, 14-31 March 1994.

Bogstad, B. & H. Gjøsæter 1994. A method for calculating the consumption of capelin by cod in the Barents Sea. ICES J. mar. Sci., 51: 273-280.

Bogstad, B. , K. Hiis Hauge & Ø. Ulltang 1995a. Results of simulation studies using a multispecies model for the Barents Sea (MULTSPEC). Int. Whal. Comm. SC/47/ NA1. 33 pp.

Bogstad, B., K. Hiis Hauge & Ø. Ulltang 1995b. MULTSPEC - a multispecies model for fish and mammals in the Barents Sea. Paper presented to the NAFO/ICES symposium «The role of marine mammals in the Ecosystem», Dartmouth, Nova Scotia, Canada, 6-8 September 1995. 47 pp.

Bogstad, B. & S. Mehl 1996. Interactions Between Cod (*Gadus morhua*) and Its Prey Species in the Barents Sea. International Symposium on the Role of Forage Fish in Marine Ecosystems. Anchorage, Alaska, 13-16 November 1996. 23 pp.

Bogstad, B. & S. Tjelmeland 1995. A method for estimation of predation mortalities on capelin using a cod-capelin model for the Barents Sea. Working Document to the ICES MSAWG, Bergen, 21-28 June 1995.

Dommasnes, A. & K. Hiis Hauge1994. HERMOD - a single species model for the Norwegian Spring Spawning herring stock. ICES C.M. 1994/H:11, 12 pp.

dos Santos J. & M. Jobling 1995. Test of a food consumption model for the Atlantic cod. ICES J. mar. Sci. 52: 209-219.

Huse, G. and R. Toresen 1995. Predation by juvenile herring (*Clupea harengus* L.) on Barents Sea capelin (*Mallotus villosus*) larvae. Pp. 59-73 in Hylen, A. (ed.) 1995. Proceedings of the sixth IMR-PINRO symposium, Bergen, 14-17 June 1994. Institute of Marine Research, Bergen, Norway.

Jobling, M. 1988. A review of the Physiological and Nutritional Energetics of Cod, Gadus Morhua L., with particular reference to growth under farmed conditions. Aquac. 70: 1-19.

Mehl, S. & N.A. Yaragina 1992. Methods and results in the joint PINRO-IMR stomach sampling program. Pp. 5-16 in B. Bogstad & S. Tjelmeland (ed): Interrelations between fish populations in the Barents Sea. Proc. fifth PINRO-IMR Symp. Murmansk, 12-16 August 1991. Institute of Marine Research, Bergen, Norway.

Tjelmeland, S. 1995. The dependence of management strategies for Barents Sea cod and capelin on the recruitment models. Pp. 305-323 in Hylen, A. (ed.) 1995. Proceedings of the sixth IMR-PINRO symposium, Bergen, 14-17 June 1994. Institute of Marine Research, Bergen, Norway.

Tjelmeland, S. 1996. Documentation of the AGGMULT model. http-address available from the author.

Tjelmeland, S. & B. Bogstad 1993. The Barents Sea capelin stock collapse: A lesson to learn. Pp. 127-139 in S.J.Smith, J.J.Hunt & D.Rivard (ed.) Risk evaluation and biological reference points for fisheries management. Can. Spec. Publ. Fish. Aquat. Sci. 120.

System Model (Systmod) for the Norwegian Sea and the Barents Sea

By Johannes Hamre[1] and Einar Hatlebakk[2].
[1] Institute of Marine Research, Box 1870 Nordnes, Bergen, Norway.
[2] Norwegian Government Computing Centre, Box 6664 Rodeløkka, N-0502 Oslo, Norway.

PROJECT DESCRIPTION

Introduction and background

The significance of stock interactions and climate change for the ecological balance in the Barents Sea was visualised in the mid-1980's when an acute deficit of prey species led to mass death among the predators. The crisis occurred after a long period of overfishing the stocks of herring and capelin, and was triggered by a climatic change to favourable recruitment conditions for herring and cod. However, the herring stock was too small to exploit the possibilities, while the cod stock became too large in relation to its food supply. Given the lack of juvenile herring and a reduced capelin stock, the rapidly growing cod stock grazed down all other available prey species in the area, including its own progeny, and starved cod, seabirds and seals appeared on the north Norwegian coast. The capelin fishery collapsed, and the traditional coastal cod fisheries were struck by the most serious crisis on record.

The stock evolution in the 1980's is the background for the development of "Systmod" the system model for the fisheries in the Norwegian and Barents Seas (Hamre et al. 1993). The model was developed in co-operation between The Institute of Marine Research (IMR) and Norsk Regnesentral (Norwegian Computing Center). So far it includes herring, capelin and cod.

The importance of stock interactions for fishery management is also the background for the IMR's MULTSPEC model. The philosophy behind MULTSPEC is to build a multispecies model based on fundamental biological and oceanographic explanatory variables. The MULTSPEC approach of model construction (the "bottom-up" approach) aims at modelling the basic mechanisms of the system. This aim is ambitious, and the work has necessarily to be carried out in a long term perspective.

An alternative to this strategy is to consider the system on a superior level, the so-called "top-down" strategy, which merely aims at modelling the most essential components of the system from the top of the food-chain and downwards. Such models may take into consideration the most vital dynamic of the key stocks only, the interaction between them and the influence of the environment. The algorithms of the models may be formulated as simple as possible since the intention is to establish a framework of the system, including the interaction between the main components only. The models may be suitable for simulating long term development, as historic data very often exist in an aggregated form.

"Systmod" should represent status of the knowledge on stock interactions which are operational at this stage. The model is an expansion of the single species management models for herring and capelin, and is basically built upon these. The model parameters are estimated on the basis of relevant data and on a priori biological knowledge.

The conceptual hypothesis of this model are described by Hamre (1988, 1994) and presuppose that the herring has a decisive impact on the dynamics of the Barents Sea ecosystem. Herring and capelin are the predominant plankton feeders in the Norwegian Sea and Barents Sea, with cod being the predominant predator (the fisheries have in periods dominated the mortality of all stocks in the model). The parameters of recruitment and growth are related to climatic changes in the Barents Sea. A warm climate provides good recruitment and growth of all species, but strong year classes of herring in the Barents Sea entails mass death of capelin fry. These ecological interrelations are the most important dynamic elements in the system, which "Systmod" is supposed to simulate. The historical data and evidences which support these hypothesis are summarised below.

The environmental conditions

The physical conditions in the Norwegian Sea - Barents Sea region are governed by the inflow of Atlantic water through the Faroe-Shetland Channel. Two main branches of the Atlantic Current create two separate ecosystems, one in the North Sea and one in the Norwegian Sea - Barents Sea.

The strength and properties of the inflowing Atlantic water vary periodically and determine the climate of the region. Mean temperature in the Barents Sea in the period 1900-1990 are shown in Figure 1. The strength of inflowing Atlantic water shows abrupt increases in the mid 1900, late 1910th, late 1930s, 1940s, 1950s, late 1960s and early 1970s, early 1980s and 1990. These periods of warm climate coincide with strong year classes of herring and cod (see Stock and Climate Interrelationship).

The ecosystem

The rich plankton production in the Norwegian Sea and Barents Sea has been harvested by two large stocks, the adult Norwegian spring-spawning herring *Clupea harengus*, and the capelin, *Mallotus villosus*. The capelin is the main plankton feeder in the Barents Sea, but in years with high inflow of Atlantic water and strong herring year classes the juvenile herring plays an important part as prey species in the area. They are the main food sources for a large variety of fish stocks, mammals and birds. The Northeast Arctic cod, *Gadus morhua*, is the largest predator in the system and plays a decisive role in the balance of the predator-prey relationship. All the main stocks of fish and mammals are harvested by man.

Relevant features of the general biology of the fish stocks are illustrated in Figure 2. Prior to the 1970s, the adult herring fed in the Polar front area of the Norwegian Sea during summer and autumn, but spawned on the Norwegian west coast in the early

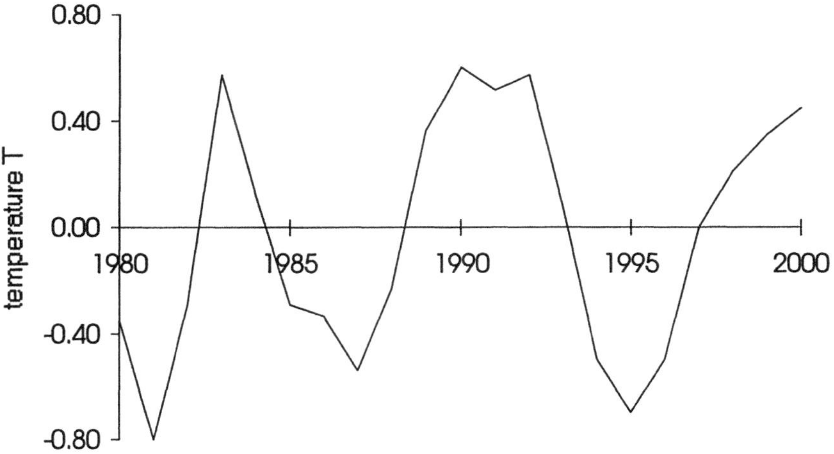

Figure 1. Mean annual temperature in the Barents Sea (Kola section), in 1900-1994 (smoothed over 3 years).

spring. The migration transferred huge quantities of fish biomass from distant waters to the Norwegian coast and turned the Norwegian west coast into one of the richest fishing and hunting grounds in the world. Young herring (0-4 years old) are widespread on the Norwegian continental shelf and in the southern parts of the Barents Sea, when the inflow of Atlantic water is high. In the latter area the young herring distribution overlaps that of juvenile capelin and may affect recruitment to the capelin stock. Due to heavy exploitation, the herring stock collapsed in 1969. The feeding migration to the Norwegian Sea was disrupted and has not yet been resumed. The stock has in later years been distributed in Norwegian coastal waters throughout the year.

The capelin spends its whole life in the Barents Sea. During summer and autumn the cacheable stock (2-4 years old) feeds in the marginal ice zone of the northern Barents Sea, and the maturing capelin migrates to the coast of Norway and Murmansk for spawning during winter and early spring. The spawning migration of capelin has the same positive effect on the Norwegian north coast fisheries as the herring had for the fisheries on the west coast. The capelin suffers mass mortality after spawning, and the life span of the fish is therefore relatively short, 3-5 years (Prokorov, 1965).The larval and juvenile stages are distributed in the central and south-eastern Barents Sea, where they are subjected to predation by juvenile herring and cod.

In addition the areas are inhabited by two large semipelagic stocks, the blue whiting and the polar cod. The semipelagic stocks are however of marginal importance as prey species because they spawn in other areas.

The spawning grounds of Arctic cod are strategically situated on the central Nor-

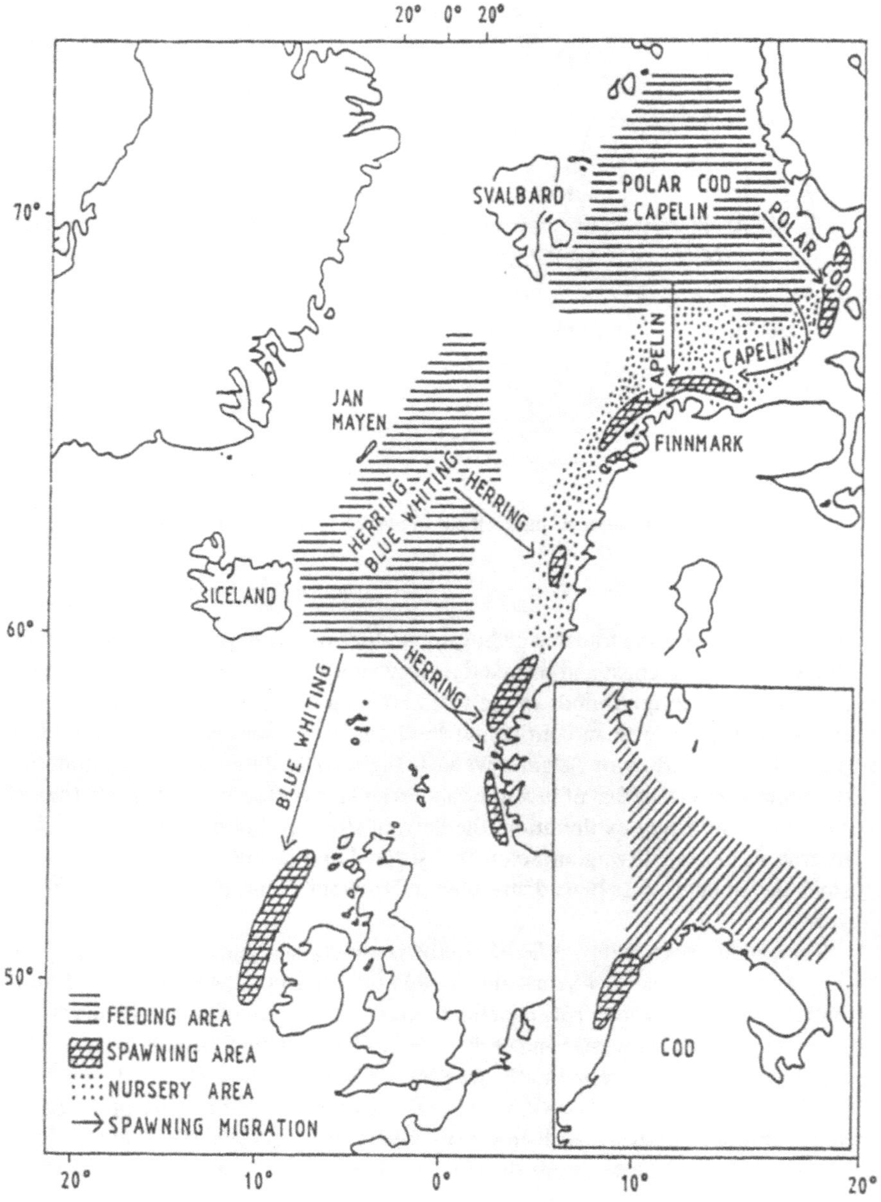

Figure 2. Distribution and migration of the most important fish stocks in the Norwegian-Barents Sea ecosystem.

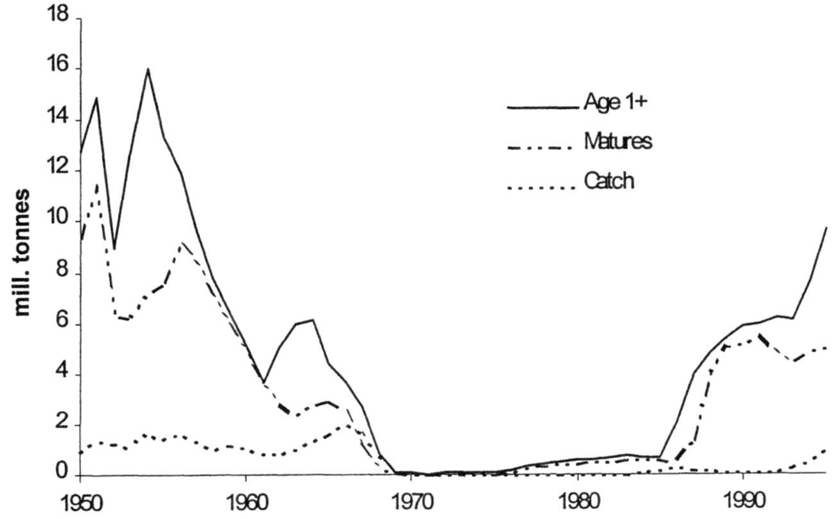

Figure 3. Estimated stock biomass of 1 year old and older herring (solid line), spawning stock biomass (broken line) and catches (dotted line) in the years 1950-1993.

wegian coast between the spawning grounds of the herring and the capelin. The cod feeds in the southern and central part of the Barents Sea and in the Svalbard area, but the distribution is mostly limited to the area south and west of the marginal ice zone. The adult capelin is therefore mostly available as food for cod during the capelin's spawning migration in winter and spring, whereas the cod may feed on juvenile herring and capelin throughout the year (Ponomarenko et al., 1975; Mehl, 1989).

Stock abundance and exploitation

Herring

VPA-based (virtual population analysis) estimates of the state of stock and the catches from 1950 onwards are summarised in Figure 3. The adult herring stock was estimated to be 7 to 10 million tonnes in the 1950s, but declined to a level of some 3 million tonnes in the early 1960s. This was mainly due to poor recruitment in the cold period 1951-58. Two strong year classes were recruited in 1959-60, but the stock was depleted in the late 1960s due to increased exploitation (Dragesund et al., 1980). A fraction of the 1969 year class survived and gave rise to a new spawning stock of herring from 1973 onwards. In the mid 1970s the spawning stock was estimated to be 100000 tonnes, and increased slowly to some 500000 tonnes in 1983. In 1983-85 three relatively strong year classes were recruited as 0-group in the Barents Sea. These year classes (especially 1984-85) suffered heavy predation from an increasing stock

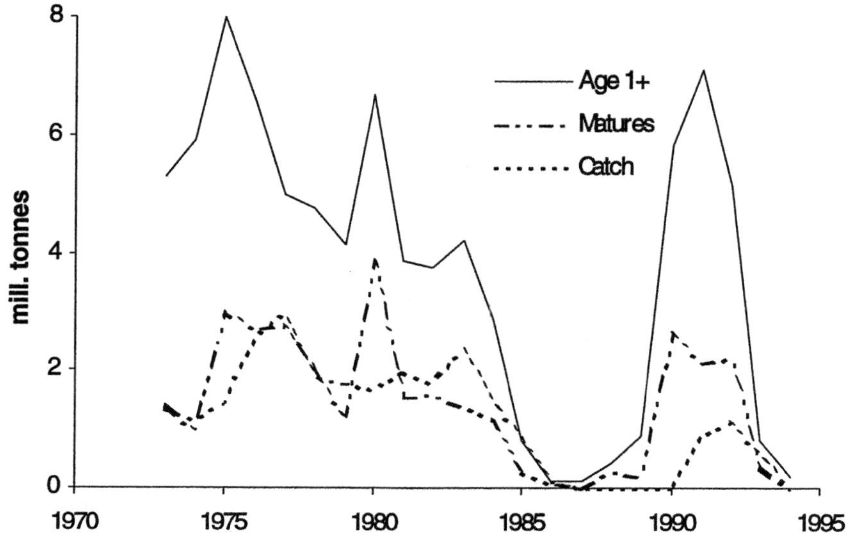

Figure 4. Estimated stock biomass of 1 year old and older capelin (solid line), maturing stock biomass in September (broken line) and catches (dotted line) in the years 1972-93.

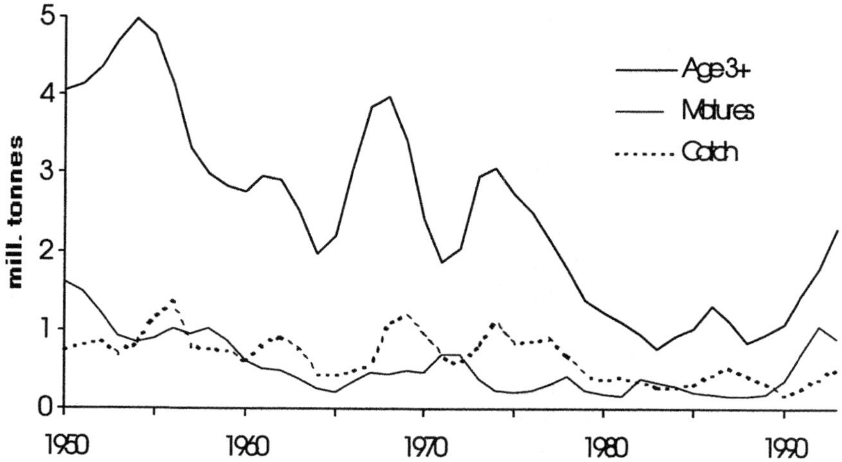

Figure 5. Estimated stock biomass of 3 years old and older cod (solid line), the spawning stock biomass (broken line) and catches in the years 1950-1993.

of juvenile cod (Mehl, 1989). The 1983 year class matured in 1988, and the spawning stock increased to some 2 million tonnes.

The recruitment is variable, and very strong year classes in relation to parent stock have occurred in 1950, 1959-60, 1963 and 1983. A new period of favourable recruitment conditions started in 1989 and the subsequent herring year classes are relative abundant, especially the 1992-year class. The fishery has been regulated by catch quotas since the early 1970s.

Capelin

Prior to 1960s, the capelin was regarded as an important stock as food for cod, but not as a large potential resource for exploitation. The capelin was fished on a small scale as prespawners off Finnmark since the 1950s. The catch increased in 1959-61, but a sudden collapse of the fishery occurred in 1962, because the year classes 1960-61 being extremely poor compared with the previous ones (Olsen, 1968). This coincides with the occurrence of the abundant herring year class 1959. From 1965 onwards the catch increased substantially and reached a record catch of 2.9 million tonnes in 1977 (Fig. 4). These large catches also include immature capelin which has been fished in the feeding area during summer and autumn since 1968. Since 1978 the fishery has been regulated by separate catch quotas for the autumn and winter fishery. From the autumn 1986 onwards the capelin fishery was banned, but the fishery was reopened in 1991. (Anon., 1993; Hamre, 1985).

Acoustic estimates of stock abundance are available since 1972 (Fig. 4). The estimates (solid line) were obtained in September and cover the age groups 1 years old and older (Dommasnes and Røttingen, 1985; Anon., 1993). The maturing of the capelin is correlated to the size, the capelin above 14 cm in September is expected to mature and spawn in the subsequent winter (Hamre and Tjelmeland, 1982). The latter estimate adjusted by the natural mortality and the catch in October-April is taken as the stock biomass of spawners in the subsequent year (broken line).

The capelin stock continued to increase in the first half of the 1970s and reached a maximum in 1975 of about 8 million tonnes. In this period no herring occurred and strong capelin year classes were recruited. These year classes matured and spawned in 1976-1977, and the stock declined accordingly. This decline in stock size was caused by increased spawning mortality mainly and was little influenced by the fishery. The natural mortality of immature capelin (M) has been estimated to 0.61 in 1974-78, 0.86 in 1979-83 and 1.68 in 1984-85 (Anon., 1987). The dramatic increase in M in latter years is caused by predation associated with an increase in the stock of young cod from 1983 onwards (Mehl, 1989). In 1984 and 1985 the recruitment also failed, probably due to increased predation or food competition between capelin larvae and the juvenile herring, and the capelin stock collapsed in 1986. The evidence thus suggests that the capelin fishery and predation by cod contributed to some extent to the stock decline in the 1980s, but recruitment failure, caused by the juvenile herring was the main reason for the stock collapse in 1986 and in 1962 as well. No collapse in the capelin stock occurred in the 1970's when the herring was absent, although the catch of capelin and the stock of cod were large.

The year classes after 1983 are weak, until the 1989 year class, which was extraordi-

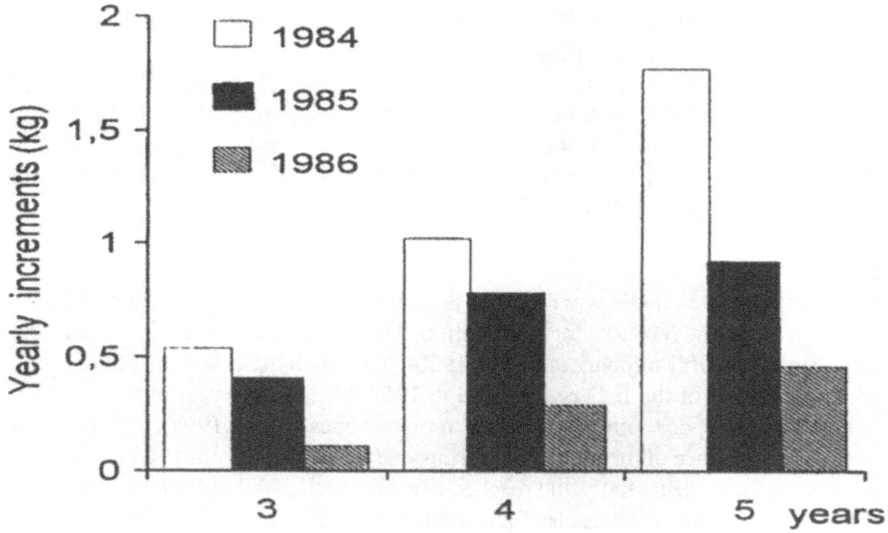

Figure 6. Yearly weight increment of cod by age in 1984-1986.

nary abundant in relation to the parent stock. In 1989-90 the individual growth of capelin increased to the highest level recorded, and the stock biomass rose from less than 1 million tonnes in 1989 to 6 million tonnes in 1990. A similar abrupt fall in the stock size has taken place in recent years (Fig. 4). The stock has suffered from recruitment failure since 1991, and a pronounced increase in M since 1992. This coincides with the occurrence of abundant year classes of herring and cod (Anon., 1993, Anon., 1994).

Cod

VPA-based estimates of the state of the stock from 1950 onwards and yearly catches are illustrated in Figure 5. In the middle of the 1950s a very strong 1950 year class was recruited, and the stock increased to about 4.5 million tonnes. In the subsequent years the stock declined to 2 million tonnes by 1964. This reduction was mainly due to increased exploitation. In later years the variation in stock size reflects to a large extent the variation in year class strength. Strong year classes were recruited in the early 1970s and the stock increased to 3 million tonnes in 1975. This is an important observation because it shows that strong year classes of cod do not prevent good recruitment of capelin. In the cold period after 1975 the recruitment was poor, and the exploitation high, and the stock declined to 1 million tonnes in 1983. From 1982 onwards several abundant year classes were recruited, but the stock biomass did not develop as expected. Shortage of suitable food since 1985 caused a dramatic fall in the individual growth rate (Fig. 6), and information on cod food in subsequent years indicate high mortality due to cannibalism (Mehl, 1989). This resulted in a dramatic fall in stock number from 1986 to 1988 (Anon., 1988). Although the exploitation of all

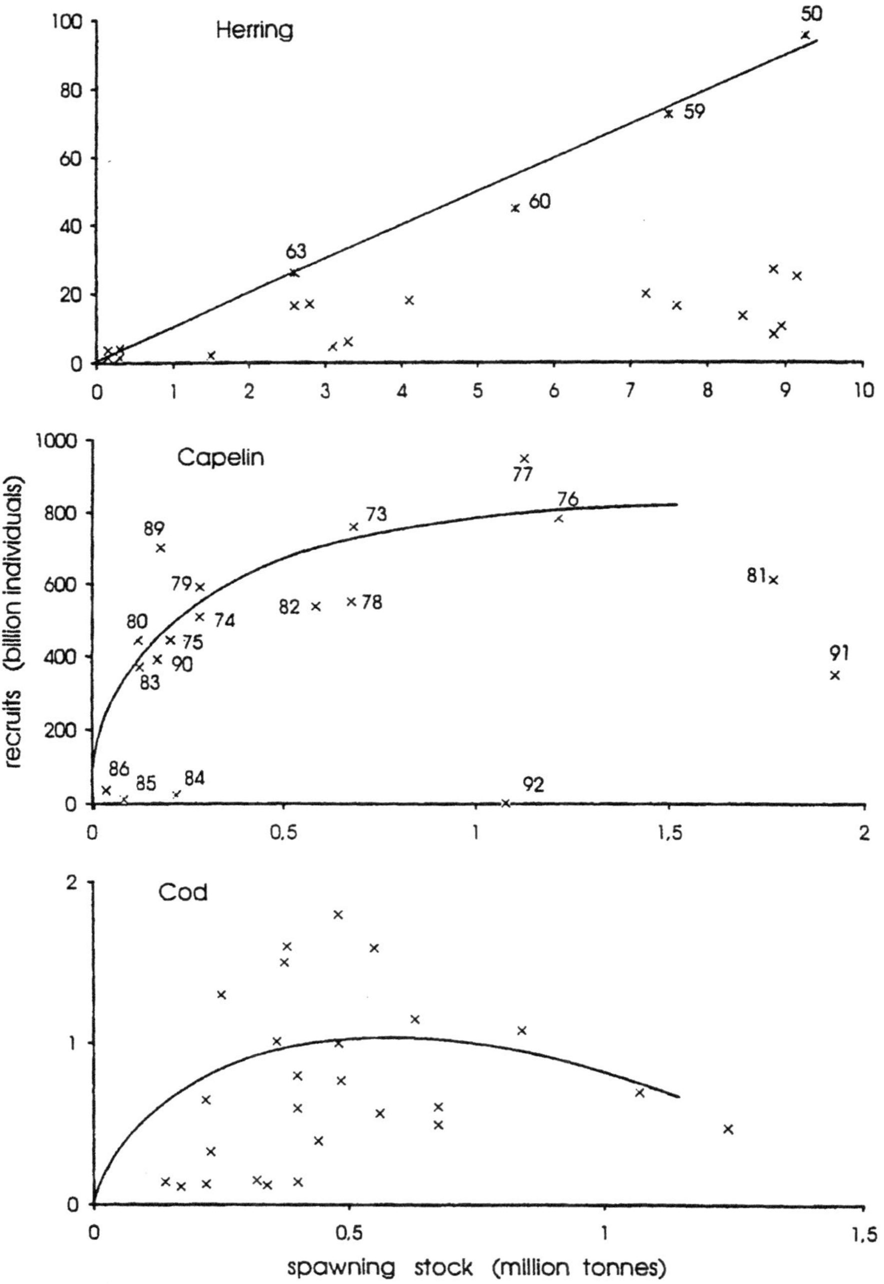

Figure 7. Stock versus recruitment plots of herring, capelin and cod. Recruits of herring and capelin are estimated as 1 year old, cod as 3 years old. The numbers indicate year classes (50 = 1950 year class).

age-groups above 3 years has been high since the early 1960s, the fall in stock size was thus a result of food shortage more than an effect of the high exploitation rate. The year classes recruited after 1989 are abundant and the stock biomass has increased to nearly 3 million tonnes (Anon., 1994).

Mammals and birds

Mammals and birds are important predators on plankton feeders and fish-feeding fishes in the Barents Sea, but data on stock sizes and feeding behaviour are inadequate to assess how much they consume. Their effect on the mortality rate is therefore considered as a part of the general natural mortality rate.

Recruitment mechanism

Plots of recruitment versus spawning stock of herring and cod for the years 1950-1980 are shown in Figure 7. The recruitment figures are back-calculated by VPA; increased mortality on juvenile herring in warm periods, when juvenile cod is abundant, is not accounted for by VPA. Both the peak herring year classes, and those recruited immediately after the peaks, are therefore likely to be underestimates. The latter may originally have been strong but may have, as in 1984-85, served as food resource for a growing stock of young cod.

The stock/recruitment function of herring suggests that in periods with favourable recruitment conditions, the strength of the year classes is proportional to the size of the spawning stock. The supply of juvenile herring as food for cod may thus be proportional to the abundance of the adult herring stock, whereas the cod may recruit strong year classes on relatively low stock levels. This means that in a state of reduced stocks, for instance after a long period of cold climate and/or heavy exploitation, a latent imbalance state of the predator-prey relationship may develop, and will be triggered when a shift to a warm climate occurs. In the early 1980s, when the climate changed, the cod stock was large enough to recruit strong year classes. Conversely the herring, which needs a large parent stock to produce abundant year classes, could not take full advantage of the improved recruitment conditions, and an abrupt shortage of cod prey occurred. These recruitment mechanisms thus explain the paradox that the crisis in the fisheries in the mid 1980's developed after a change in the climate which favours recruitment conditions for the most important species of the system.

The capelin recruitment versus spawning stock plots (Fig. 7) are closely related to a Beverton/Holt recruitment curve, except for the years 1984-85 and 1992 when the recruitment failed completely, probably due to the recruitment of abundant herring year classes.

Stocks and climate interrelationship

Information on the variation in the climate and the age structure of the stocks of cod and herring is available since the beginning of this century. This indicates that periods with increasing temperatures and a warm climate favour recruitment of cod and herring, whereas a cold climate is associated with low recruitment (Marty and Federov, 1963; Sætersdal and Loeng, 1984). Extraordinary strong year classes of cod and herring occurred around 1904, 1918, 1937, 1950, 1959 and 1963 which coincide with peaks in the temperature curve of Figure 1. In the warm period of the early 1970s, several abundant year classes of cod were recruited, but the herring failed to recruit due to the depletion of the spawning stock. The outstanding year classes of capelin and cod in the early 1970s suggests that only strong herring year classes are responsible for mass mortality of capelin fry.

Conclusions relevant to the structure of "Systmod"

In conclusion, the evidence indicates that the herring is the key prey species at fish level of the food chain in the Norwegian Sea - Barents Sea ecosystem, and the cod is the dominant predator. The capelin has an opportunistic life pattern, whose prospects of survival are governed by the state of the stocks of young herring and cod. The abundance of immature herring determines the survival of 0-group capelin, whereas the abundance of cod and death after spawning, determine the mortality of older capelin. The dynamics of the system are governed by the inflow of Atlantic water, which determines distribution, recruitment success and growth of the main species involved. Based on this knowledge, the structure and interrelationships of the system are modelled as follows:

1. Herring, capelin and cod are explicitly modelled as age- and length distributed, and recruitment, growth and mortality are parameter dependent.
2. The recruitment is modelled as stock-dependent but the initial recruitment success is determined by the climate. Increasing temperature in the Barents Sea favour good recruitment, whereas decreasing temperature may result in recruitment failure for herring and cod.
3. The growth of herring and capelin is made density dependent and dependent on the changes in the climate. The growth of the cod is made dependent on the abundance of juvenile herring and the state of the capelin stock.
4. Stock interrelationships are modelled as mortality rate. The survival of 1-group capelin is linked to the predation rate of juvenile herring in the Barents Sea, whereas the survival of older capelin depends on the catch and the predation rate of cod. Mortality rate of capelin postspawners is modelled separately (spawning mortality). Survival of juvenile herring in the Barents Sea is made dependent of predation of cod. Survival of 1-group cod is depending on the stock abundance of juvenile cod and modelled as cannibalism, when other prey species are scarce.
5. The periodical variation in ocean climate is modelled as mean temperature anomalies in the Barents Sea, following a sine curve.
6. The distribution of stocks is not modelled by areas but specific parameters for mature and immature fish are given. As the distribution patterns of mature and immature fish differ, the stock components will indirectly be split into areas. The most important interactions related to overlap in the distribution of stocks can thus be taken care of by the model.
7. The interaction with fishing activity is modelled as fishing mortality acting simultaneously with the predation and the natural mortality rate.

MODEL STRUCTURE

"Systmod" includes the species capelin, herring and cod in the Barents Sea and in the Norwegian Sea. The stock is divided into age groups, and the immature and mature part of the stock. The juvenile herring stock is also divided into two parts, the part of the stock which feeds in the Barents Sea and the stock component feeding on the Norwegian coast.

The stocks in number are divided into age and length groups. For each length group of juvenile fish the mean weight is constant, whereas the weight of mature fish is reduced after spawning according to a specified value. The weight of a length group will gradually increase to the initial value until the next spawning time.

The stocks of herring and cod are read from the data files 1 January, and the capelin is reprocessed from the stock estimate in October the previous year. The recruitment is computed and added to the stocks, and the mortality and growth throughout the year are computed on a monthly basis. This is the main dynamics of the yearly stock progress, which is repeated.

Growth

The model is length based, and the individual growth is simulated by an upwards movement of the individuals on the length scale. The growth in length is modelled with the following equation (von Bertalanffy, 1938):

$$dL(t) = l(t+T) - L(t) = (L_{\infty} - L(t)) \cdot (1 - e^{-KT}) \cdot M(t)$$

where t is a time variable and T an interval of fixed length. L is the maximum length of the fish. M(t) is a factor which distributes the yearly growth on the different months. K is the growth parameter which determines the growth related to the size of the stock and environmental factors (capelin and herring). K is determined by the following equations:

$$K = (a + b \cdot e^{-d \cdot B(t)}) \cdot g$$

B(t) is the stock abundance at time t. The exponential term regulates the densitydependent growth. For cod the parameter K is computed according to the equation:

$$K = (a + b \cdot \frac{CCOD}{COD}) \cdot g$$

CCOD is the food consumption for cod, and COD the stock biomass. g is a function to regulate the growth according to environmental factors, where T is the deviation in temperature from the mean (see Temperature model):

$$g = a + bT$$

The biomass is computed from the mean weight of each length group. The maximum biomass production for each stock may be limited by a parameter, and from this parameter the maximum value of K is computed. All variation in growth is modelled by the growth equation of length increment.

The relation between length and weight are expressed by the relation:

$$W = c \cdot L^3$$

where W is weight, L is length, and c is a constant called condition factor. This formula is used when reading stock data to interpolate between length groups where there is no weight data.

Temperature model

Historical data shows that the climate in the Barents Sea affects the recruitment and growth of cod, herring and capelin (see Stocs and Climate Interrelationships). There seems to be 8-12 years between warm periods in the Barents Sea, and the temperature is modelled by a sine function:

$$\Delta T(t) = p_2 \cdot \sin(\frac{2\pi t}{p_1})$$

The temperature (T) is generated as deviation from "normal" by parameters for amplitude (deviation from mean), frequency and start year. When connected to modules for recruitment and growth, the impact on the stocks of periodic changes in the climate is computed.

Recruitment

The juveniles are recruited to the stock in January as age 1, or about 9 months old. To describe the relation between the spawning stock in the springtime and the number of recruits in January the following year, the Beverton-Holt function is used:

$$R = \frac{M \cdot B}{H + B} e^{a \cdot T}$$

where R = recruitment; M = maximum recruitment; B = spawning stock biomass; H = the half value; T = temperature deviation; a is a parameter.

The exponential term in the equation favours recruitment when T is high. Warm climate in the Barents Sea increases the stock of juvenile herring, which affects the survival of capelin fry (see Recruitment Mechanism). This is modelled by a reduction factor proportional to the strength of the age groups 1 to 3. In this reduction factor also 1-group cod, 2-group and 3+ capelin are included:

$$R = R_0 \cdot (1 - (a_1 \cdot HER_1 + a_2 HER_2 + a_3 \cdot HER_3 + a_4 \cdot COD_1 + a_5 \cdot CAP_2 + a_6 CAP_{3+}))$$

In the equation HER_1 denotes 1-group herring, HER_2 2-years herring, HER_3 3 years herring, COD_1 denotes 1-year old cod and CAP_2 and CAP_{3+} denote 2 years and older capelin respectively.

Maturation

Maturation is modelled as a pulse. All the species mature at the turn of the year. The maturity ogive (fractions of mature) in each length group is computed from a logistic function (Tjelmeland, 1987):

$$M(l) = \frac{1}{1 + e^{4P_1(P_2 - l)}}$$

where l denotes mean length in the length group (midpoint in length interval). The two values P_1 and P_2 are parameters which have a relevant biological interpretation, since we have:

$$M(P_2) = 1/2$$

P_2 is the length where 50 % is expected to mature, while P_1 is the ratio number for the number of mature individuals at this length.

$$M'(P_2) = P_1$$

Mortality

In "Systmod" there are four types of mortality:

- spawning mortality
- fishing mortality
- predation
- natural mortality (not including predation)

The predation includes the species which are part of the model, but will mainly be cod. Natural mortality includes mortality caused by all other reasons.

In computing mortality the following variables are used:

$$N_{i,t+1} = N_{i,t} \cdot (1 - G) \cdot (1 - M) \cdot (1 - P_i) \cdot (1 - F_i)$$

Spawning mortality (G)

Spawning mortality is modelled for the capelin stock only. A parameter in the model determines the fraction of mature capelin which survives after spawning (1. April).

Predation (P)

Predation is an important component of the model. It is computed each month and has impact on the growth of cod and the mortality of the prey species. The cod is predator for capelin and juvenile herring and also includes cod cannibalism.

The monthly mortality rate caused by the predation is computed by species.

For capelin by the formula: $P_i = K \cdot (a_0 \cdot COD_{im} + a_1 \cdot COD_m)$
For Barents Sea herring: $P_i = K \cdot COD/(1 + b_1 \cdot CAP)$

For cod: $P_i = K \cdot COD/(1 + b_1 \cdot CAP + b_2 \cdot HER_{im})$

P_i – predation mortality rate for length group i
K – K is a constant
a_i(COD) – weighted sum of the predator stock (m/im = mature/immature)
b_i(CAP, HER) – reduction in mortality rate due to preference of prey species

Natural mortality (M)
The natural mortality excluding the predation is modelled by a constant mortality factor (M) which reduces the stock in number each month.

Fishing mortality (F)
The fishing mortality is modelled by a constant factor (F), but limited by setting quotas for seasons and years. The fishing mortality is acting until the quota is reached.
The fishing pattern is modelled as a catchability factor specified for each length group. The probability that a fish is caught is the catchability factor multiplied by the relative number of fish in the length group concerned. The catchability factor is set to 0 for fish which is shorter than a given length, and for fish bigger than another length, the catchability is set to 1. The catchability is increased gradually between 0 and 1.

DATA

Input data
Input data for "Systmod" are stock data and parameter data files. Stock data are needed to describe the state when the simulation begins. The stock data are also used for comparing the simulated data with the real stock estimate for the same period. Parameter data are needed for the algorithms which compute the development of the stocks during the simulation period.
The stock data files contain data for each species which is included in the simulation. The files contain yearly stock estimates, and the estimates are divided into age and length groups. For each group mean weight, number and maturation per cent are given.
The stock data for cod and herring are VPA estimates. In order to distribute the age data from the VPA into length groups, a standard deviation is set for each age group. Cod data are available from 1946 and herring data from 1950. The capelin data are aggregated data from a database. Capelin data are available from 1972, as acoustic stock estimate obtained in October each year.
There is one parameter file with control variables, and another file which contains the stock dependent variables. The control variables are stock independent parameters. They determine the methods that will be used, the interactions between the species, the temperature and how much output there will be from the simulation.

In the file with the stock dependent parameters there are one set of parameters for each species. There are parameters for the different components of the model as recruitment, growth, maturation, fishing mortality and natural mortality.

Output data

The main report from the simulation gives the total biomass production and mortality for all the species. There is one line for each year, and in the first column the stock size is given. The rest of the columns give the main numbers for the simulation the last year. The biomass production is divided in recruitment and growth. The total mortality is divided into fishing mortality, spawning mortality, predation and other natural mortality.

There are several output files which are intended to be used when the model is parameterised. To study the recruitment there is a file which describes the recruitment for selected years and the development the next two years. There is a file which describes the development of selected cohorts with length, weight and number at different age. It is useful when setting parameters for the growth. There is also a file which describes the mortality for one year. Numbers for the different mortality types are given.

A stock file is written at end of the simulation with the state of all the species in the simulation. The file is written in the same format as the input stock files. It is also possible to write the state of the stocks the last 12 months of the simulation to file.

PARAMETER ESTIMATION

Procedure of estimation

The model parameters are estimated by comparing model results to data in the period 1982 to 1993. The influence of the climate on growth and mortality (see Growth and Recruitment) is modelled by the variable T derived from observed temperature anomalies in the Kolo section (Fig. 8). For prognostic purposes the development in the climate is modelled as described in the Temperature model section.

The general principle of parameter estimation has been a step-by-step procedure, picking groups of parameters which are supposed to be correlated, and arrange them in a hierarchy according to their importance for the results. This screening of parameters has mainly been based on biological knowledge of the system and experience gained in the model runs.

Ranging of stocks has also turned out to be of importance in reducing the complexity of the estimation problem. In the step-by-step procedure one should preferably start with the herring parameters, because they are relatively little affected by the other stocks (predation by cod on juveniles only). The dynamics of the capelin stock are strongly influenced by the two other species, and the capelin parameters have therefore been estimated at the end of the procedure.

The output files of model results compared to data are transferred to a spread sheet for graphic processing.

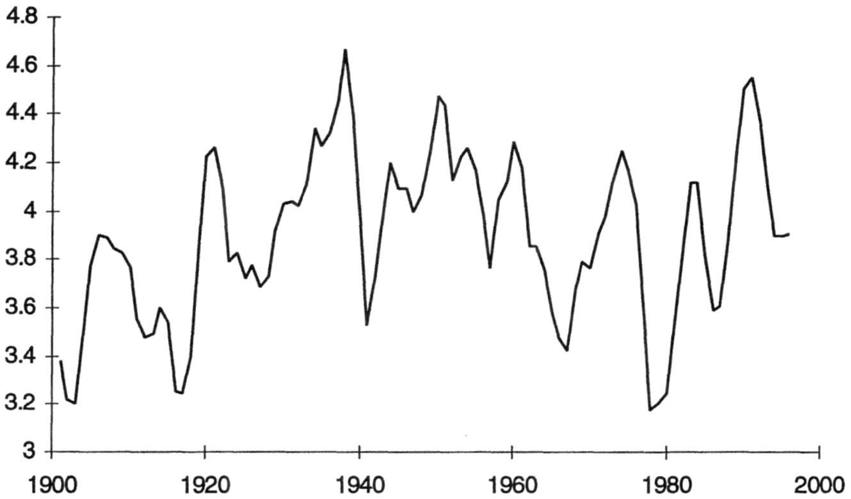

Figure 8. Annual deviation from the mean temperature observed at the Kola section in the Barents Sea in the years 1980-1994. From 1995 onwards modelled values as described in the Temperature model section.

Recruitment, growth and mortality estimates

The effect of climate changes on recruitment and the species interaction between capelin and juvenile herring, i.e. the parameters a, a_1, a_2, a_3 (in the Recruitment section) are the most decisive parameters in modelling the dynamics of the system. The predation of cod on capelin and herring is also important and has a significant effect on the biomass production of the system. This predation effects the mortality of capelin, the size of the juvenile herring stock which regulates the recruitment of capelin, and the recruitment to the adult herring stock. The total consumption (CCOD) determines the growth of cod (see Growth). The first approach to fit the model estimates to data has therefore been concerned with the adjustments of these parameters mainly. The majority of parameters, determining details of stock structures and exploitation have been judged according to general knowledge of stocks and fisheries and model results have so far not been tested against data.

The recruitment parameters of capelin are determined by comparing modelled number of 2 years old in October to acoustic estimates of the age group in autumn. Such estimates are available since 1972. The recruitment or survival of 0-group capelin is strongly effected by the juvenile herring stock, and provision for cod predation and for cannibalism is also made in the reduction factor in the Recruitment section. Although no reference on capelin cannibalism is known, this is the most reasonable way to model the extraordinary high recruitment of capelin observed in 1989. By assuming that the maximum recruitment M in the stock-recruitment formula is large,

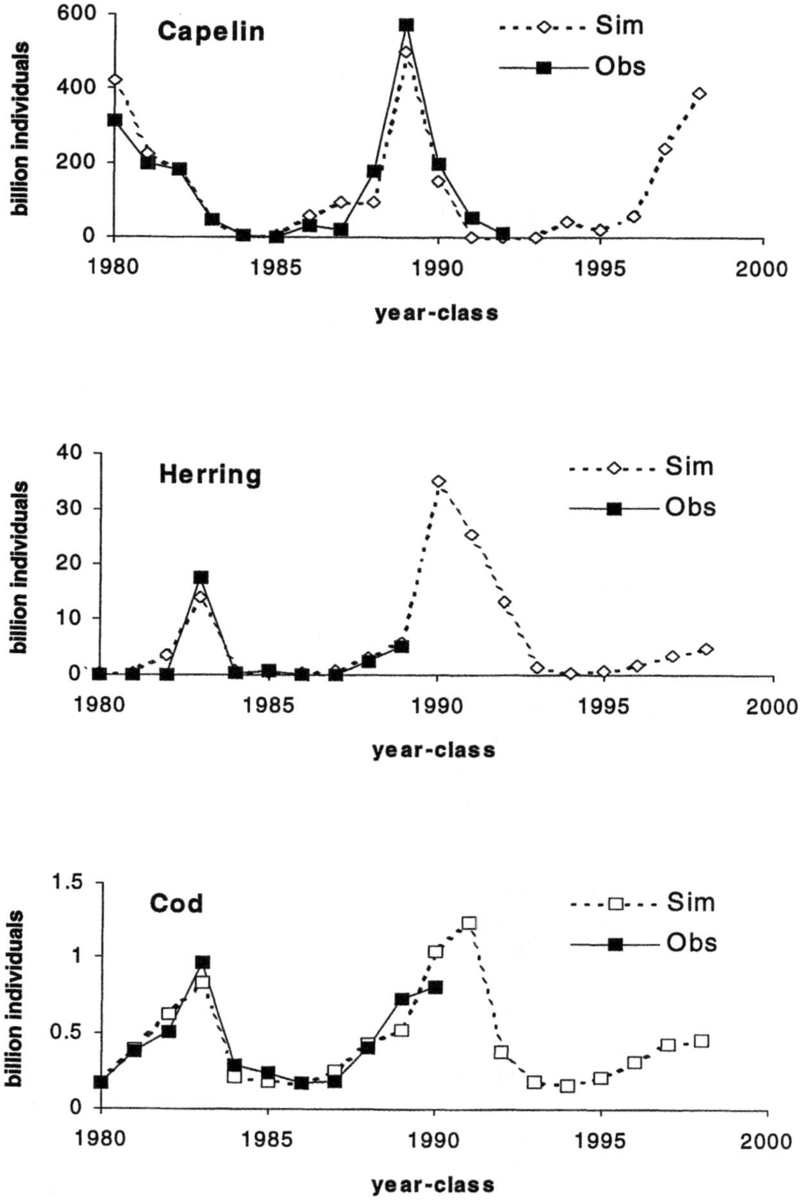

Figure 9. Simulated and observed recruitment by year class of capelin, herring and cod. The estimated capelin year classes refer to 2 years old, for herring and cod to 3 years old fish.

but that the effect is reduced by some form of cannibalism when the stock is large, the high recruitment potential of capelin will be active when the stocks of herring and capelin are small, as in 1989. This mechanism may explain the extraordinary quick rebuilding of the capelin stock at the end of the 1980s (Fig. 10).

The recruitment parameters of herring and cod are determined by comparing modelled number of 3 years old to the corresponding VPA-estimates of the stocks. The estimates are depending on the temperature, both through the exponential factor of the stock recruitment function and by selecting higher maximum recruitment level (M) when the temperature increases above a given value. The survival of 3 years old herring and cod is more over determined by predation of immature cod. The model results compared to data are shown in Figure 9.

The recruitment estimates are the basis for modelling the stock in number, which is a function of recruitment and the total mortality rate. The total mortality is modelled by the historic catch, the predation mortality and the mortality due to other causes (see Mortality). The predation parameter of cod on capelin is determined by comparing the model results to estimates of yearly consumption of capelin by cod (Bogstad and Mehl, 1995). The results are shown in Figure 11. The parameter of herring consumption by cod is judged on the basis of information derived from Barros and Toresen, 1995. The modelled stocks in number by year compared to data are shown in Figure 10.

The conversion of stock in number to stock biomass is modelled by the growth equation of length increment and a factor of weight by length group (see Growth). The growth of capelin and herring is made temperature-dependent whereas the variation in growth of cod depends on the consumed biomass of prey species only. The model results compared to data are shown to the left in Figure 10.

The model results fit reasonably well to the corresponding data observed. This support the underlying theory of the dynamics of the system, i.e. that the recruitment and growth are governed by the ocean climate and that the stock interrelationship determines the relative abundance of the three main species. In the 1970s when the herring stock was depleted and absent from the Barents Sea, the cod fed on capelin mainly and although the capelin stock was heavily fished, no serious food-shortage for cod did occur. This implies that the Barents Sea cod can manage without the herring or even obtain a more stable food supply when the herring is absent. In Figure 12 is shown the simulated development of capelin and cod in the 1980s when the herring is removed. All the other parameters including the catches are unchanged. In that case the model simulates a temporary reduction in the capelin stock in the middle of the 1980s, but the stock is rebuilt to a level of 3 mill. tonnes in 1986. The cod stock is not effected to any appreciable extent in number, but the biomass increases gradually throughout the period. The model thus supports the hypothesis that it was the appearance of a reduced stock of herring which caused the crises in the cod fishery in the 1980s. The herring was abundant enough to prevent capelin recruitment, but the stock was not sufficiently large to feed the cod (Hamre, 1994). This example

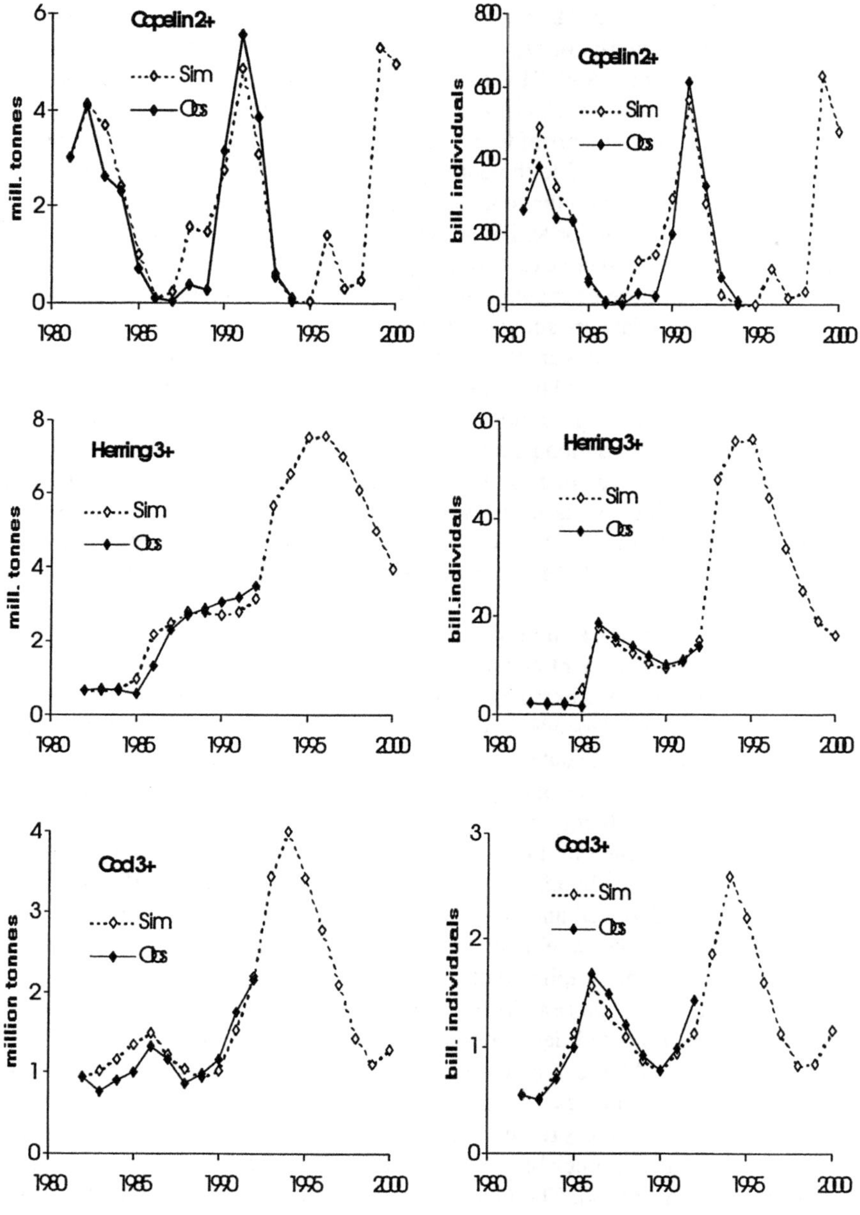

Figure 10. Simulated and observed stocks of capelin, herring and cod. The estimates of capelin include 2 years old older (2+), for herring and cod 3 years and older (3+).

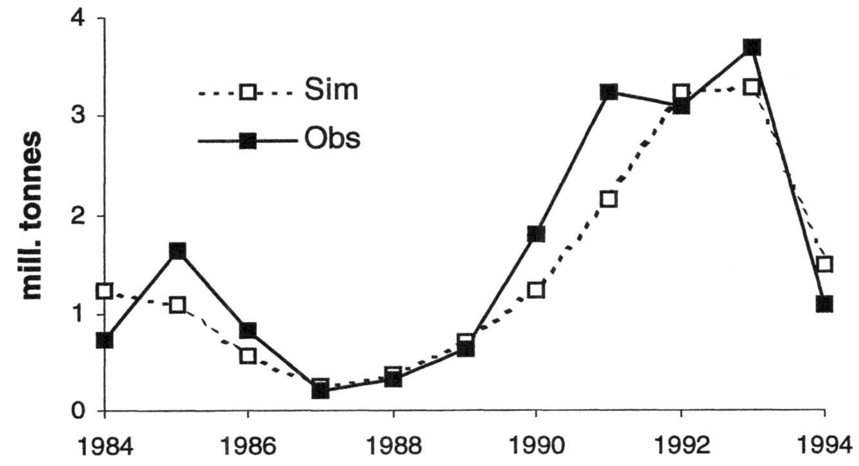

Figure 11. Estimated and simulated consumption of capelin by cod in the years 1984 to 1994.

shows more over that the model simulates an acceptable relationship between climate changes and stock interaction also in the case when the herring stock is depleted and absent from the Barents Sea.

APPLICATION OF THE MODEL

When all the model parameters are estimated and adjusted to available data, the model may be used in fishery strategy analysis and in catch and stock forecast. An example of the latter use are included in Figures 9 and 10. The basic assumption of the forecast is that the temperature in 1994-2000 will change as simulated by the temperature model (see Temperature model) and shown in Figure 8. The yearly catch from 1996 onwards is set to 1.2, 0,8 and 0 million tonnes for herring, cod and capelin respectively. A cold period in the middle of the 1990s is thus assumed to give poor recruitment in the first coming years, but a new warm period will occur at the end of this century. This may open for capelin recruitment from 1996 onward, and a new pick in the capelin stock will occur at the end of the 90's. The stocks of herring and cod increase in 1994-1996, but due to poor recruitment after 1992, the stocks will again be reduced to low levels at the end of the 1990s. It should, however, be stressed that this simulated stock development is not supposed to be a prognosis of the future prospects of the fishery, but is only an example to show how the model may be used in a simple projection of stocks and yield.

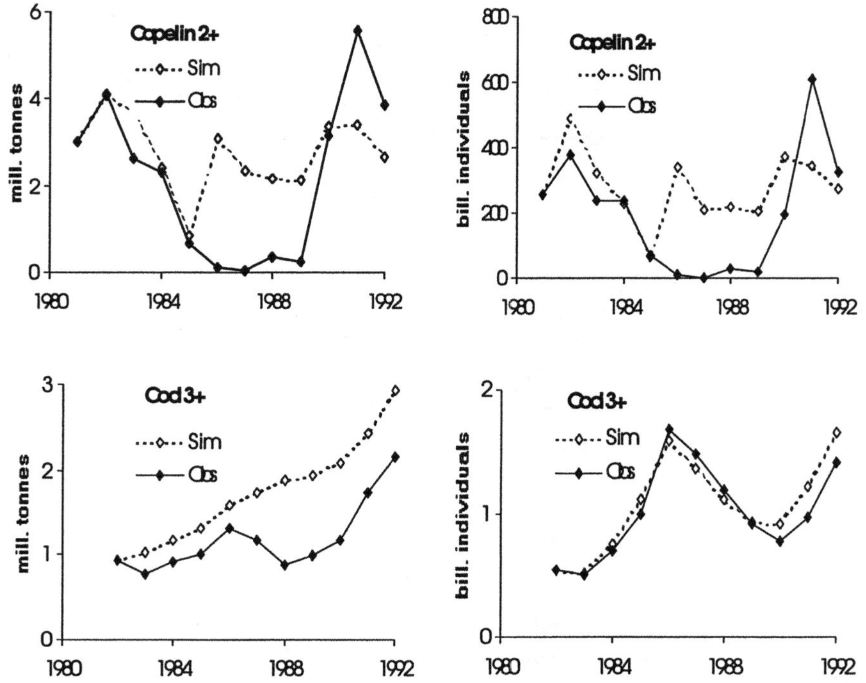

Figure 12. Simulated stocks of capelin and cod when herring is removed from the model, compared to observed data. Symbols as in Figure 10.

REFERENCES

Anon., 1987. Report of the Atlanto-Scandian Herring and Capelin Working Group, Copenhagen 27-31 October 1986. ICES C. M. 1987/Assess:8.

Anon., 1988. Report of the Arctic Fisheries Working Group, Copenhagen 14-24 September 1987. ICES C. M. 1988/ Assess:5.

Anon., 1993. Report of the Atlanto-Scandian Herring and Capelin Working Group, Copenhagen 19-23 October 1992. ICES C. M. 1993/Assess:6.

Anon., 1994. Report of the Arctic Fisheries Working Group, Copenhagen 24 August -2 September 1993. ICES C. M. 1994/Assess:2.

Barros, P. & R. Toresen 1995. Modelling age dependent natural mortality of juvenile Norwegian spring-spawning herring (*Clupea harengus* L.) in the Barents Sea. Pp. 243-261 in Hylen, A. (ed.): Precition and relevance of pre-recruit studies of fishery management related to fish stocks in the Barents Sea and adjacent waters. Proc. sixth IMR-PINRO symp. Bergen, 14-17 June 1994. Institute of Marine Research, Bergen, Norway.

Bogstad, B. & S. Mehl 1995. The diet of North-East Arctic cod 1984-1994. Working document to the Atlanto-Scandian Herring, Capelin and Blue Whiting Working Group, October 1995.

Dommasnes, A. & I. Røttingen 1985. Acoustic Stock measurements of the Barents Sea capelin 1972-1984. A Review. Pp. 5-24 in: Gjøsæter, H. (ed.). Proc. Soviet-Norwegian Symp. Barents Sea Capelin, Bergen-Norway, 14-17 Aug. 1984. Institute of Marine Research, Bergen.

Dragesund, O., J. Hamre & Ø. Ulltang, 1980. Biology and population dynamics of the Norwegian spring spawning herring. Rapp. P-v.Reun.Cons.int.Explor.Mer., 177: 43-71.

Hamre, J. 1994. Biodiversity and exploitation of the main fish stocks in the Norwegian Sea - Barents Sea ecosystem. Biod. Cons. 3 (1994): 273-292.

Hamre, J., E. Hatlebakk, G. Hagen, K. Merckoll, 1993. System-modell for havfiske, NR-notat, STAT/5/1993: 1-25.

Hamre, J. 1988. Some aspects of the interrelation between the herring in the Norwegian Sea and the stocks of capelin and cod in the Barents Sea. ICES C.M. 1988/H: 42: 1-15. (Mimeo).

Hamre, J. 1985. Assessment and management of Barents Sea capelin. Pp.5-24 in: Gjøsæter, H. (ed.). Proc. Soviet-Norwegian Symp. Barents Sea Capelin, Bergen-Norway, 14-17 August 1984. Institute of Marine Research, Bergen.

Hamre, J. & S. Tjelmeland 1982. Sustainable yield estimates of the Barents Sea capelin stock. ICES C.M. 1982/H: 45: 1-24. (Mimeo).

Marty, J.J. & S.S. Fedorov 1963. Features of the population dynamics of marine herring as seen from the Atlanto-Scandian stock. Rapp.P.-v.Reun.Cons.perm.int.Explor.Mer, 154: 91-97.

Mehl, S. 1989. The Northeast Arctic cod stocks consumption of commercially exploited prey species in 1984-1986. Rapp.P.-v.Reun.Cons.int.Explor.Mer, 188: 185-209.

Olsen, S. 1968. Some results of the Norwegian capelin investigations 1960-1965. Rapp. P.-v-Reun. Cons. int. Explor. Mer, 158: 18-23.

Ponomarenko, V.P. & I.Y. Ponomarenko 1975. Consumption of the Barents Sea capelin by cod and haddock. ICES C.M. 1975/F: 10: 1-6. (Mimeo).

Prokhorov, V.S. 1965. Ecology of the Barents Sea capelin (*Mallotus Villosus* (Müller)) and prospects for its commercialutilization. Fish. Res. Bd. Can. Trans. Ser., No.813: 1-131.

Sætersdal, G. & H. Loeng 1984. Ecological adaption of reproduction in Arctic Cod. Pp.13-35. in: Proc. of the Soviet-Norwegian Symp. Reproduction and Recruitment of Arctic Cod. Inst. of Mar. Research, Bergen, 1984.

Tjelmeland, S. 1987. Mortality on Barents Sea Capelin calculated by the IMR Multispecies Model for the Barents Sea. ICES C.M. 1987/H: 55: 1-21 (Mimeo).

"Stockfish", a Multispecies Model for Stochastic Analysis

By Erling Moxnes*.
SNF, Breiviken 2, N-5035 Bergen-Sandviken, Norway.

INTRODUCTION

We explore the possibilities for constructing an aggregate model of the multispecies fish resources of the Barents Sea, useful for the purpose of policy analysis under uncertainty. A wish to make the model compatible with existing Norwegian models of the Barents Sea, notably the Aggmult and the Econmult models, works to complicate the model. A wish to use the model to search for policies under uncertainty, using computer intensive search procedures, calls for a limited number of calculations per simulation. Focus here is on the biological part of the model. The economic part of the model and the stochastics are presented in Moxnes (1994).

In the next section the model is discussed in light of existing literature. An effort has been made to enable direct use of prior information when formulating the model. In the Empirical investigation section some of the assumptions are seen in light of data for the fish resources of the Barents Sea. An attempt is made to explain recruitment for some of the species in terms of biomass measures of the other species as well as ocean temperature. For capelin in particular, the randomness is significantly reduced. Finally in "Comparison to historical development", the model is simulated with a suitability matrix derived from a version of Aggmult. Comparisons with historical data indicate that there is ample room for improvements of the model and its parameters. The lowest age classes represent the weakest part of the model. In spite of its weaknesses, the model is close to a stage where it can be used for policy analysis under uncertainty. In particular, the model could be used to test the appropriateness of further simplifications, to see if the complexity of current multispecies models is warranted.

---

* Thanks to Sigurd Tjelmeland and his colleges at the Institute of Marine Research in Bergen for much useful information and helpful comments, thanks to numerous other informants mentioned in footnotes here and in the original report.

THE "STOCKFISH" MODEL

The model has been designed with two main constraints in mind: It should resemble the multispecies models already being used in Norway, Aggmult (Bogstad et al. (1992)) and Econmult (Eide (1991)), and it should be simple. Besides transparency, the latter constraint follows from the need to perform computer intensive search for optimal policies. The major simplification compared to Aggmult is the choice of a pure biomass model where numbers, length and weight are not modelled explicitly. Furthermore, parameters are aggregated over seasons. The major simplification compared to Econmult is that the fishing fleet is aggregated into two homogenous fleets only. Stockfish is programmed in Fortran and makes use of Excel to present results.

The biological model has 3 main species: capelin, herring, and cod. There are 6 age classes for capelin, 11 for herring and cod. In addition the model has a simplified representation of plankton and sea mammals, with one "age class" each.

Since the initial letters of cod and capelin are the same, we resort to the scientific names for distinguishable acronyms for the three main species.

$$s = \begin{cases} p - \text{plankton} \\ v - \text{capelin} \quad (\textit{Mallotus villosus}) \\ h - \text{herring} \quad (\textit{Clupea harengus}) \\ m - \text{cod} \quad (\textit{Gadus morhua}) \\ w - \text{sea mammals} \quad (\text{seals and whales}) \end{cases}$$

Biomass, B_{sa}, for each species, s, and age class, a, is given as:

$$\frac{dB_{sa}}{dt} = P^g_{sa}/\alpha - P^L_{sa} - (c_{sa} + s_{sa} + m_{sa})B_{sa} - H_{sa} \tag{1}$$

where P^g_{sa}/α is the gain in biomass from predation on others. Here, P^g_{sa} is the biomass that is eaten, while α measures the amount of food to be eaten for each unit of food that is utilised. (James and Stark (1982) models defecation as an explicit loss term). Loss by predation is measured by P^L_{sa}. Natural mortality (normally measured as a fraction of the number of fish) is captured by m_{sa}, while c_{sa} is a measure of the biomass that is needed for catabolism, the part of metabolism which gives energy to keep the fish alive and moving (normally measured in terms of a fraction of the weight). Biomass loss due to spawning is captured by s_{sa}[1]. Fishing mortality is given by H_{sa}.

The model interactions, notably predation, are characterised by a large number of feedback loops with non-linear relationships. It is impossible to find an exact discrete time representation of such a system where parameter definitions are maintained.

[1] The factors c_{sa} and s_{sa} are related to the parameter c_{10} in the weight equation in Bogstad, et al. (1992). Our expression is chosen because it allows direct use of prior information.

Hence a discrete model cannot benefit directly from prior information about relationships and parameters. Therefore a continuous representation has been chosen. A first order Euler approximation is used to discretize the model to carry out simulations. The time step is set short enough to obtain satisfactory accuracy.

Once a year, all biomasses are shifted up one year class, see the first line in equation 2. Fish moving up the age classes behave according to a new set of parameters each year. Within each year, the parameters stay constant.

$$B_{sa;t} = \begin{cases} B_{s;a-1;t-1} & \text{normally when a} \geq 0 \\ R_{s;t} & \text{normally when a} = 0 \\ B_{sa;t-1;} & \text{s = p and s = w when a} = 0 \\ B_{sa;t-1;} + B_{s;a-1;t-1} & \text{s = h when a} = 10 \end{cases} \quad (2)$$

In the zero classes, a=0, the entering biomasses equal recruitment, R_S, for the fish species. For plankton and sea mammals, the biomass is just replaced by itself. In the ultimate year class for herring, a=10, the biomass of the lower age class is added to the remaining biomass of the ultimate year class. The reason for this is that infrequent recruitment to the herring stock might at times imply that recruitment becomes dependent on classes of age higher than 10 years.

Sea mammals and plankton are treated different from the other species. The growth of plankton, see equation 1a, is described by the surplus growth curve in the first expression on the right hand side. A major reason for this choice is the short lifetime of plankton. Carrying capacity for plankton is denoted by B_{p0cc}, and the maximum growth is given by g_{p0}. As in equation 1, P^L_{p0} denotes predation loss.

$$\frac{dB_{p0}}{dt} = 4g_{p0}\left(1 - \frac{B_{p0}}{B_{p0cc}}\right)\frac{B_{p0}}{B_{p0cc}} - P^L_{p0} - d_{op}\frac{B_{p0}}{\tau_r} + \max\left[0, R_{p0} - \eta_p\left(T - T_m\right)\right] \quad (1a)$$

Due to ocean currents, some plankton will flow out of the Barents Sea. If the plankton density was the same all over the sea, outflow would be given by the plankton biomass divided by the residency time of the water, τ_r. However it seems likely that the density drops off towards the exit line. The coefficient d_{op} denotes the relative plankton density of the outflowing water. The equation also allows for an exogenous inflow of plankton to the Barents Sea, «recruitment» R_{p0}. This inflow is argued to be negatively correlated with ocean temperature. The mean temperature, T_m, is subtracted from the temperature, T, such that R_{p0} represents the expected inflow. The max-function ensures that the inflow stays non-negative.

For sea mammals the biomass is simply assumed to stay constant. The slow dynamics of sea mammals implies that the assumption is acceptable as long as food supplies for sea mammals are not close to the subsistence level.

$$\frac{dB_{w0}}{dt} = 0 \quad (1b)$$

Predation gain by species, s, and age class, a, is given as:

$$P_{sa}^{g} = \frac{\phi_{sa}}{\phi_{sa} + \phi_{rs}} \alpha (g_{sa} + c_{sa} + s_{sa}) B_{sa} \tag{3}$$

The term α (g_{sa}+c_{sa}+s_{sa}) B_s denotes maximum predation. The maximum predation is needed to produce maximum net growth, g_{sa}, catabolism, c_{sa}, and spawning, s_{sa}, of the biomass B_{sa}. α makes up for a limited utilisation of the food intake. The term ϕ_{sa}/ (ϕ_{sa}+ϕ_{rs}) denotes the fraction of the maximum predation the fish will catch. This fraction depends on the likelihood of finding food, or the food abundance, ϕ_{sa}. This is called the feeding level concept by Andersen and Ursin (1977), feeding rate by James and Stark (1982) and functional response by Holling (1965). The feeding level concept is used in the Aggmult and Multspec models, Bogstad, et al. (1992). The parameter ϕ_{rs} denotes the food abundance for which the food uptake is half the maximum uptake. In James and Stark (1982), ϕ_{rs} depends on the biomass of competing predators. Like most other modellers we ignore this direct effect of competition and assume that competition is captured indirectly through the effects on the biomass of prey populations.

Food abundance, ϕ_{sa}, is given by the availability of prey species, S, and prey age classes, A. For each prey species and age group, SA, a matrix β measures its suitability as food for the predator species and age groups, sa. The suitability matrix captures average effects of geographical distribution and seasonal migrations, catchability, as well as preferences by the predator.

$$\phi_{sa} = \sum_{SA} \beta_{saSA} B_{SA} \tag{4}$$

Predation loss for species, S, and age class, A, is given by the sum of predation by other species and age classes, sa. The fraction of the total predation, P^{g}_{sa}, that is directed towards SA, is given by the ratio of food of type SA (β_{saSA} B_{SA}) to the total food abundance ϕ_{sa}. In other words, predators are assumed to be opportunistic feeders, eating what is most easily available at the time, given the weights imposed by the suitability matrix.

$$P_{SA}^{L} = \sum_{sa} P_{sa}^{g} \frac{\beta_{saSA} B_{SA}}{\phi_{sa}} \tag{5}$$

The spawning fraction, s_{sa}, represents an average for males and females. Its distribution over age classes reflects the maturity of the fish. Average maturation time and the distribution of maturity over age classes are kept constant by the formulation. James and Stark (1982) have chosen a similar approach: «an amount of biomass equal to some proportion of the average age-class biomass over the preceding month is 'sacrificed' in reproduction».

$$S_s = \sum_a s_{sa} B_{sa} \tag{6}$$

It seems rather straight forward to get approximate estimates for spawning as a fraction of the biomass. However, what determines the mortality of the zero class is not well known. There seems to be non-linearity's involved that are not easily explained by predation from higher age classes. Therefore we use a nonlinearity to convert spawning into recruitment. A Beverton-Holt recruitment function is chosen, where spawning replaces mature biomass. Since spawning is a fraction of the biomass, this variation has no practical effect as long as the half-value, S_{rs}, also refers to spawning products. R_{sm} is the maximum recruitment.

$$R_s = R_{ms} \frac{S_s}{S_s + S_{rs}} \tag{7}$$

Harvesting is modelled much the same way as predation. Total harvesting of species, s, is determined by catch per unit effort, X_s, utilisation, U_s, fleet size, N_f, and a standard for harvest per boat, H_{bf}.

$$H_s = X_s H_{bf} U_s N_f \tag{8}$$

Catch per unit effort is modelled as a functional response, where ϕ_{hs} is the abundance of species s, and ϕ_{hrs} is the half-value.

$$X_s = \frac{\phi_{hs}}{\phi_{hs} + \phi_{hrs}} \tag{9}$$

The abundance of each species is given by the biomass vector, B_{sa}, multiplied by a vector for fleet selectivity, β_{hsa}. Selectivity of the fleets corresponds to suitability for predation.

$$\phi_{hs} = \sum_a \beta_{hsa} B_{sa} \tag{10}$$

The harvesting of each age class, a, for species, s, is given by the total harvest of the species, H_s, times the ratio of the abundance of the age class, $\beta_{hsA} B_{sA}$, to the total abundance ϕ_{hs}.

$$H_{sa} = H_s \frac{\beta_{hsA} B_{sA}}{\phi_{hs}} \tag{11}$$

This concludes the presentation of the model.

How does the model capture different biological phenomena? Geographical distribution is not explicitly modelled. The geographical distribution of the species is reflected in the coefficients of the suitability matrix. Where there is no overlap between age classes of different species, suitability is zero. The selectivity vectors for harvesting also reflect the geographical distribution.

Seasonal migration of species is not captured. The quarterly suitability matrices used in the Aggmult model have been averaged to a yearly matrix in Stockfish. Similarly, selectivity vectors represent yearly averages.

Numbers, length and weight of the fish is not explicitly modelled, the model only operates with biomass. Similar approaches have been taken in the multispecies and multicohort models by James and Stark (1982) and Larkin and Gazey (1982). Pure biomass models are known from the Lotka-Volterra equations, and from single species models of the Schaefer type. Pure biomass models are also known from multicohort models, see e.g. Spulber (1985). Several Nordic models explicitly model numbers, length and weight of the fish, notably the Multspec and the Aggmult models, see Bogstad, et al. (1992), models by Hatlebakk et al. (1993), and by Sparre and Willmann (1993).

The major reason for using a pure biomass model is simplicity, three sets of state variables are replaced by one set. A potential problem with biomass models is that they might sustain biomass even in the absence of recruitment; then the weight component of growth makes up for lacking numbers. This problem is avoided in Stockfish by the shift function, see equation 2. If no biomass enters the zero class, there will be no biomass entering higher age classes after a while. Since Stockfish operates with highly uncertain recruitment, it is important to avoid this problem.

Some weaknesses of the biomass formulation that are not avoided, have to do with short run dynamics. In a "number, length and weight reality" it is likely that weight will vary more strongly with food abundance than the number and the length of fish. This implies that the maximal growth rate of a year class will depend on the conditions of the fish in addition to its age, given that a starved fish can grow more quickly than a fat one. In Stockfish the parameters for maximal growth will represent an average fish condition. Consequently, Stockfish might come to underestimate growth when a fish population is starved, and overestimate growth when hunger is satisfied. The effect is probably of minor importance.

Another dynamic effect concerns spawning. Maturation depends on the length of the fish, which in turn depends on the food supply. A young fish will not spawn, no matter how fat it is, if it is not long enough. Hence, Stockfish will not be able to distinguish a meagre but mature age class from a fat but immature age class. However, the problem is of limited importance because the average maturation time has a quite limited standard deviation. In Stockfish the maturation time is constant.

Ageing is captured by shifting the biomass content towards higher age classes once a year. Age classes have different coefficients for growth (predation, catabolism and mortality) and recruitment (spawning). Ageing ensures that several biological phenomena are captured.

Over the lifetime of the fish, body growth dominates the lowest age classes while natural and fishing mortality dominates the highest age classes. Thus, biomass will reach a maximum for some medium age class. This is where the harvest should be targeted to maximise yield per recruit.

If recruitment is a limiting factor, harvesting must be such that a sufficient amount of fish are allowed to reach mature age classes. Hence, the cohort model allows for the stock-recruitment perspective. Of particular importance in a model with uncertainty, is the possibility that the entire recruitment for a species comes to depend on one particular age class. This is more or less the case for herring. Then policies should ensure that this vital age class is not decimated through fishing.

Representing age classes will contribute to dynamic behaviour that could not come out of aggregate models. Most important is the time lag between recruitment and mature fish. Of importance is also the interaction between age classes of the same species, cannibalism. This seems to be a likely explanatory factor for observed fluctuations in recruitment of three year old cod in the Barents Sea, see Eide (1993). Similar fluctuations have been produced by the Aggmult model.

A potential problem with cohort models is that growth rates are not stock dependent, see a discussion in Spulber (1985). With increasing recruitment as a function of mature biomass, these models can produce unlimited growth. This problem is avoided in Stockfish since the cohort model for each species is embedded in a multispecies environment. When a species becomes very large, it will deplete its food resources, food abundance will decline and growth will stagnate.

From a user point of view as well as from a computational point of view, the formulation for predation is demanding. The suitability matrix measures 30 by 30, containing 900 elements. If predation could be treated in a more aggregate manner, the size would be reduced. One possible way to do this is to treat subsets of age classes of a species as an aggregate when it comes to predation. The individual age classes are maintained. Aggregate predation for each subset is distributed to the individual age classes according to relative biomasses within the subset. With three subsets for cod, capelin and herring, the suitability matrix would be reduced to 11 by 11, or 121 elements. We have not tested this simplification yet.

EMPIRICAL INVESTIGATION

The statistical philosophy applied is a crude version of Bayesian statistics. The model is constructed such that most parameters have "real-life" interpretations. This enables direct use of prior information about these parameters. Parameters, for which there are little prior information, are to some extent estimated or calibrated by the use of time-series data. See Forrester and Senge (1980) and Zellner (1981) for other elements of our model building strategy. The individual parameters are presented in Moxnes (1994). Here we report the more interesting results.

Cod's predation of capelin and plankton

In the following we use data from the Institute of Marine Research to make some rough tests of the relationships between food availability and predation. The data are measurements of food intake by cod of different prey species[2] over the years 1984 to 1992, biomass of capelin[3] for the same years, and density of plankton[4] from 1984 to 1990. Note that the data do not distinguish between age classes. First we examine a regression of predation on food availability.

Table 1 shows that there is a clear relationship between the stock size of capelin and cod predation on capelin. Furthermore, the availability of capelin explains 76 percent of the variation in the cod's predation on capelin. This result reflects the importance of capelin, which on average made up 45 of cod's total food intake over the period.

Table 2 shows that the relationship between the density of plankton and cod's intake of plankton is not as clear as the one for capelin. The relationship is significant at the 8.6 percent level, the sign is as expected, and 48 percent of the variation is

Table 1. Regression of cod's predation of capelin on biomass of capelin ($R^2 = 0.76$; N = 9).

Parameter Estimates Term	Estimate	Std Error	t-Ratio	Prob>|t|
Intercept	622	319	1.95	0.092
Biomass capelin	0.41	0.087	4.67	0.002

Table 2. Regression of cod's predation of plankton on density of plankton ($R^2 = 0.48$; N = 7).

Parameter Estimates Term	Estimate	Std Error	t-Ratio	Prob>|t|
Intercept	509	276	1.85	0.124
Density plankton	101	47	2.13	0.086

Table 3. Regression of cod's predation of capelin (per biomass cod) on biomass of capelin and density of plankton ($R^2 = 0.67$; N = 7).

Parameter Estimates Term	Estimate	Std Error	t-Ratio	Prob>|t|
Intercept	0.54	0.15	3.45	0.026
Density plankton	-0.043	0.022	-1.92	0.127
Biomass capelin	0.000013	0.00003	0.46	0.668

[2] Data are produced by Sigbjørn Mehl at the Institute of Marine Research.

[3] Data on biomass of capelin come from ICES (1992). Note that in the calculations we use the biomass estimates from the year before. This is because the capelin estimates are made in the autumn after most of the years predation has taken place.

[4] Density data for plankton from Skjoldal et al. (1992).

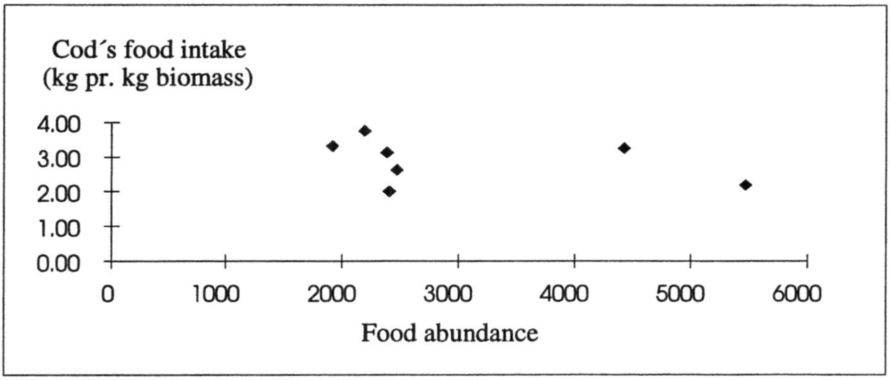

Figure 1. Observed food intake by cod for different levels of food abundance (a constructed measure).

explained. Much more should not be expected in light of imprecise measurements and definitions of plankton predation[5] and plankton density, plankton's moderate importance on the cod's menu (25 percent), and the low number of observations (7).

Going back to equation 5, we see that the way predation is modelled, predation of a species is also inversely related to the availability of the other species (through the term f_{sa}). Table 3 shows the results when cod's predation of capelin (*per biomass of cod*) is regressed on biomass of capelin and density of plankton. The signs are as expected, plankton has a negative influence on cod's predation of capelin. Furthermore 67 percent of the variation is explained. However, the results are not significant.

When looking at predation of plankton in a similar way to capelin, the same conclusions are reached, see table 4.

Further attempts to estimate parameters of the non-linear relationship in equation 3, do not lead to precise estimates of the reference food abundance. Only an upper limit is indicated, which is indicated by the data in figure 1.

Table 4. Regression of cod's predation of plankton (per biomass cod) on biomass of capelin and density of plankton ($R^2e = 0.70$; N = 7).

Parameter Estimates Term	Estimate	Std Error	t-Ratio	Prob>|t|
Intercept	0.1435086	0.14922	0.96	0.3907
Density plankton	0.0432634	0.02132	2.03	0.1123
Biomass capelin	-0.000013	0.00003	-0.48	0.6549

[5] Plankton is defined as the sum of shrimps, amphipodes and 30 percent of the category others in the data provided by Sigbjørn Mehl.

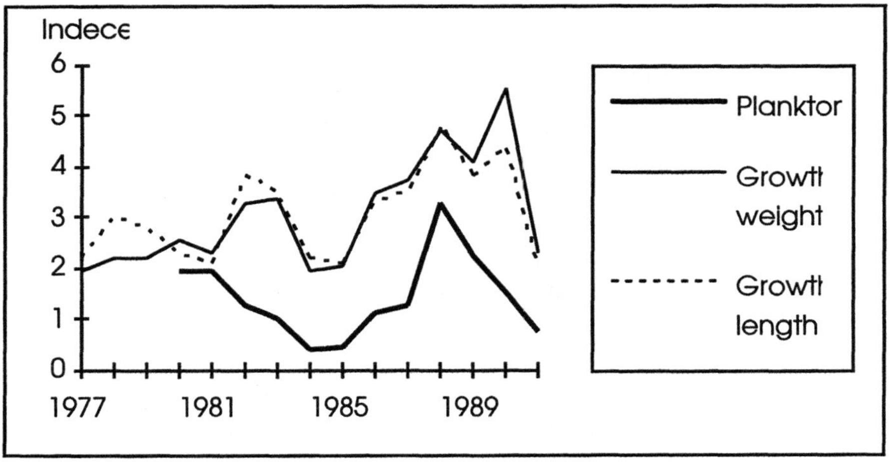

Figure 2. Capelin growth in weight and length from age one to two, source Skjoldal, et al. (1992). Plankton biomass from Moxnes (1994) using original data in Skjoldal, et al. (1992).

Capelin's predation of plankton

We do not have direct observations of predation gains of capelin and herring. However, we have some data on individual growth of young capelin (from one to two years of age). These observations of growth should be expected to be closely related to the feeding level, a relationship we will investigate.

Figure 2 shows data for growth in weight and in length of capelin. We will use the weight measure. For this choice notice that weight and length growth are highly correlated, r=0.98. This also indicates that at least for capelin, limited error will be caused by aggregating weight and length.

The regression between capelin growth and plankton density in table 5 shows that only 31 percent of the variation is explained by plankton density. The sign is as expected and the level of significance is 6 percent.

Figure 3 shows the data points together with a possible non-linear relationship. As for cod, only an upper limit for the reference food abundance is indicated by the data.

Table 5. Relationship between capelin growth and plankton density ($R^2 = 0.31$; N = 12).

Parameter Estimates Term	Estimate	Std Error	t-Ratio	Prob>|t|
Intercept	4.6498032	1.25187	3.71	0.0040
Density plankton	0.4055819	0.19217	2.11	0.0610

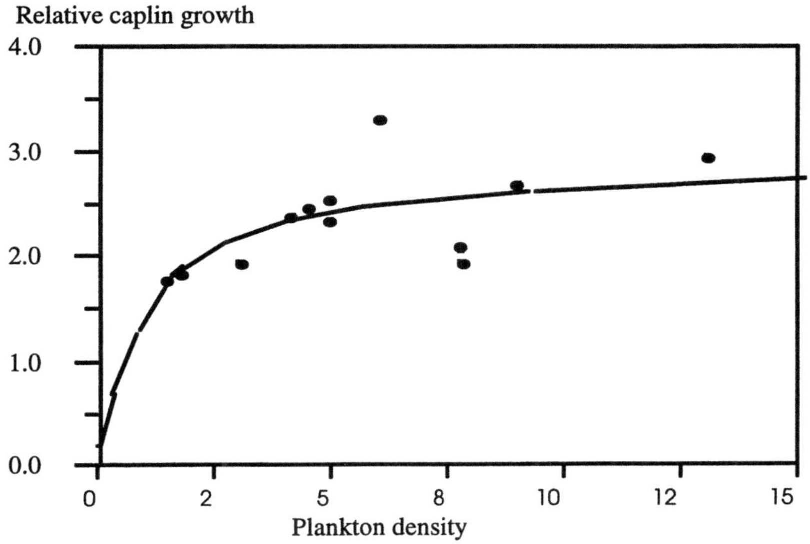

Figure 3. Relationship between relative yearly capelin growth in weight, and plankton density. The two low points are in 1980 and 1981, the high point is in 1990.

Plankton dynamics

Plankton is the main prey in the Barents Sea, it has its own dynamics, and it seems to be influenced by environmental factors. Plankton data used above, together with data for temperature[6] (September) and capelin biomass (main predator) are used to get a rough estimate of the plankton model. The estimate will be rough because the model is simplified, the data are uncertain, the sample size is small, and the observations are made only once a year and occasionally at different seasons.

The starting point for the estimation is equation 1a. Below it is written in difference form. Due to the non-linearities we use a simple first order approximation. We use a one year time step. Due to the fairly rapid dynamics of plankton, this procedure might lead to considerable errors. Predation is assumed to be given by the feeding level relationship estimated above, using a prior estimate of the reference, ϕ_{rv}=1.23, times the biomass of capelin, B_v. This product is treated as one exogenous variable when estimating parameters. The parameter a summarises the remaining parameters in the expression for predation, see equation 3. Other predators than capelin are not included.

$$B_{p0,t+1} - B_{p0,t} = 4g_{p0}\left(1 - \frac{B_{p0}}{B_{p0cc}}\right)\frac{B_{p0}}{B_{p0cc}} - aB_v \frac{B_{p0;t}}{B_{p0;t} + \phi_{rv}} - d_{op}\frac{B_{p0}}{\tau_r} + \max\left[0, R_{p0} - \eta_p(T - T_m)\right]$$

[6] Mean temperature 0-200m for the Kola section 70,30'N-72,30'N.

Since we have a density measure of plankton, we define a weighted area, A, and the equation can be rewritten in density measures, $B_{p0}=AD_{p0}$. At the same time we reorganise the equation to the form in which it is estimated (linear in compounded parameters).

$$D_{p0,t+1}-D_{p0,t}=\left(\frac{4g_{p0}}{B_{p0cc}}-\frac{d_{op}}{\tau_r}\right)D_{p0}-\frac{4Ag_{p0}}{B_{p0cc}^2}D_{p0}^2-\frac{a}{A}B_v\frac{D_{p0;t}}{D_{p0;t}+\phi_v^D}-\frac{\eta_p}{A}T+\frac{R_{p0}}{A}+\frac{\eta_p}{A}T_m$$

The parameter estimates all have the expected signs. Temperature is significant at the 3 percent level, while predation from capelin has a level of significance of 7.8 percent. The individual parameters for plankton density are far from being significant. However, the combined effect of the two plankton terms reaches a level of significance of 6.4 percent.

To get estimates for individual parameters from the compound estimates, we have to rely on prior information to reduce the number of unknowns. The outflow of plankton from the Barents Sea depends on the residency time of water in the Barents Sea, τ_r=4 years[7]. Relative plankton density of outflowing water is set lower than 1.0 according to a reasoned guess, d_{op}=0.5. Mean temperature, T_m=4.9°C. Finally, the parameter a is set equal to 6, reflecting the values for g_{va}, l_{va}, s_{va}, and α. Actually it is based on an average measure of capelin's need for food[8]. With these assumptions the parameter values shown at the end of table 6 result. Maybe surprising, the main source of plankton turns out to be inflow of plankton, R_{p0}. The maximum growth of plankton in the Barents Sea, g_{p0}, is less than one third of the expected inflow. Also

Table 6. Summary of fit for growth of plankton, $B_{p0,t}$-$B_{p0,t-1}$ (R^2 = 0.73; N = 11; Whole model, Prob>F = 0.063).

Term	Estimate	Std Error	t-Ratio	Prob>|t|
Intercept	20.6	6.63	3.10	0.021
$D_{p0;t}$	0.32	0.85	0.38	0.72
$D_{p0;t}^2$	-0.069	0.060	-1.15	0.29
T	-3.67	1.24	-2.95	0.026
$B_V\frac{D_{p0;t}}{D_{p0;t}+\phi_V^D}$	-0.86	0.40	-2.13	0.078

Individual parameter values: B_{p0cc} = 45 mill. t., g_{p0} = 5.0 mill.t./yr., R_{p0} = 17 mill.t./yr., η_p = 26 mill.t./yr./deg., A = 7.0

[7] Personal communication Hein Rune Skjoldal at the Institute of Marine Research .

[8] In a paper by Arne Hassel a capelin stock of 5 million tonns is estimated to need 30 million tonns of food, referred by Adnan Eyjad. This implies a ratio of 6.

note that the temperature dependence of the inflow rate is strong. The effect of temperature is strong enough to make the inflow of plankton negative. This happens for temperatures above 5.6 °C, of which there are two occurrences in the data used. To ensure positive inflows, either η_{p0} should be lowered or a non-linearity introduced. In the model we use a max-function, which keeps the inflow non-negative. The temperature dependence of the inflow rate raises the question whether the growth rate is also temperature dependent. The answer is thought to be yes[9]. Allowing for this effect, the results would have turned out somewhat differently. Predation from other species than capelin would also change the results.

Capelin recruitment

We start by regressing capelin recruitment on mature capelin, temperature and the predators herring and cod. Data for these variables are defined in Moxnes (1994). We use a linear model[10]. Table 7 summarises the results. The model as a whole is accepted at the 0.04 percent level, while it explains 82 percent of the variation in the recruitment data. All signs are as expected, and all parameters are significant at very low levels. A second order term for mature capelin turned out to be in-significant. Multicolinearity is not likely to be a problem[11].

When the herring in the Barents Sea is not aggregated, the distribution over the four last recruitments is shown in the lower part of table 7. The three first year classes come out as significant. Weights are declining.

Table 7. Summary of fit for recruitment of capelin (t+1), for period 1974-1989 ($R^2 = 0.82$; N = 16; Whole model, Prob>F = 0.0004; RMSE = 0.49).

Term	Estimate	Std Error	t-Ratio	Prob>|t|
Intercept	-4.23	1.14	-3.71	0.0035
Herring in Barents Sea	-1.23	0.20	-6.14	0.0001
Cod	-0.96	0.21	-4.63	0.0007
Temperature	1.47	0.25	5.90	0.0001
Mature Capelin$_{t-1}$	0.33	0.13	2.56	0.027
When splitting Herring in the Barents Sea				
Herring$_{t-3}$	-0.11	0.074	-1.52	0.1681
Herring$_{t-2}$	-0.17	0.072	-2.42	0.0416
Herring$_{t-1}$	-0.26	0.075	-3.43	0.0089
Herring$_{t}$	-0.35	0.088	-4.01	0.0039

[9] Communiation Sigurd Tjelmeland, IMR.

[10] Neither a log-linear nor a more complicated non-linear model produced a better overall fit.

[11] The correlation between temperature and herring (r=0.08 when data for 1983 is removed) and between temperature and cod (r=0.05) is very limited and of little importance for the results.

The results seem to be roughly in line with the Aggmult suitability matrix for the zero class of capelin. Cod is a significant predator. Predation from herring declines somewhat more rapidly with age classes than what is assumed in the suitability matrix.

Stability of the parameters is tested by eliminating data points, see table 8. All significant estimates except two, are close to or within the range of plus and minus one standard deviation around the original estimates. Of the two major deviations, the intercept value of -6.30 is within two standard deviations. The herring parameter of -2.70 is 7.4 standard deviations away from the original estimate. This happens when the four years around the 1983 incidence of very high herring recruitment are re-

Table 8. Parameter values for different time periods, numbers in parenthesis are not significant at the 5 percent level. Within brackets are shown the original parameter values plus and minus one standard error.

Years excluded	Intercept	Herring in B. Sea	Cod	Tempe-rature	Mature $Capelin_{t-1}$	Whole model Prob>F
None (original)	-4.23	-1.23	-0.96	1.47	0.33	0.04 %
[±std.error]	[3.1-5.37]	[1.03-1.43]	[0.75-1.17]	[1.22-1.72]	[0.2-0.46]	
1983	-3.57	-1.49	-0.96	1.36	(0.27)	0.05 %
1986-89	(-2.57)	-1.15	-0.90	1.17	(0.16)	5 %
1982-85	(-2.22)	-2.70	-0.96	1.21	(0.13)	1 %
1978-81	-6.30	-1.15	-0.75	1.84	(0.068)	0.01 %
1974-1977	-5.47	-1.31	(-0.85)	1.68	0.48	0.1 %

Table 9. Summary of fit for recruitment of capelin (t+1), no temperature and predation (R^2 = 0.003; N = 16; Whole model, Prob>F = 0.85; RMSE = 1.02).

Term	Estimate	Std Error	t-Ratio	Prob>|t|
Intercept	1.00	0.43	2.33	0.035
Mature $Capelin_{t-1}$	0.045	0.23	0.19	0.85

Table 10. Summary of fit for recruitment of capelin (t+1), plankton (t-1) added to original estimation (R^2 = 0.99; N = 10; Whole model, Prob>F = 0.0007; RMSE = 0.20).

Term	Estimate	Std Error	t-Ratio	Prob>|t|
Intercept	-6.70	1.19	-5.60	0.0050
Herring in Barents Sea	-1.23	0.11	-11.1	0.0004
Cod	-0.57	0.54	-1.06	0.3495
Temperature	1.81	0.16	11.7	0.0003
Mature $Capelin_{t-1}$	0.48	0.061	7.90	0.0014
$Plankton_{t-1}$	0.033	0.028	1.21	0.2933

moved. Estimated parameters for the effect of herring range from - 1.15 to - 2.70. In case the original estimate should have been changed, it is towards a larger absolute value. Alternatively, the results indicate that there is a non-linearity involved.

The inclusion of main predators and temperature produces a radical improvement of the rather imprecise picture one gets from simple stock-recruitment models. Table 9 shows that an attempt to explain capelin recruitment as a function of mature capelin only fails. The t-value drops from 2.56 to 0.19.

Note that temperature is the most important explanatory variable. When temperature is excluded from the regression, the R^2 drops to 0.25. Thus, even if the full model explains 82 percent of the variation, it is not very useful for predictive purposes unless precise predictions for temperature can be made. The model is more useful for the purpose of stochastic policy analysis.

Finally plankton is added as an explanatory variable to the original estimation. Only 10 data points exist with simultaneous observations of all variables involved, such that the degrees of freedom becomes only 4. Compared to the stability estimates where 13 or more data points were used, the plankton model comes out with a much lower level of significance for the whole model. Plankton gets the expected sign, although the parameter is not significant. Three parameters are highly significant (herring, temperature and mature capelin), while cod is no longer significant. All four parameters are within the confidence limits of the previous estimates. Removing temperature from this model leads to insignificant results. Hence, plankton, which is also temperature dependent, is not a good substitute for temperature in the recruitment equation for capelin.

While the determinants of capelin recruitment in the first model were predators, plankton is a prey. Since the original recruitment data measures the number of individuals, the measured effect of plankton must come through deaths. Availability of food prevents starvation and death. Higher weight of each capelin leads to increased feeding levels for predators and reduced predation in numbers. A healthy individual might also be harder to catch than a starved one. Thus, it seems likely that plankton influences capelin mortality.

Herring recruitment

Recruitment of herring is normally given by a VPA measure of 3 year olds, ICES (1992). However there also exists a logarithmic measure of larvae in the zero year class. Thus there is the possibility of first explaining the number of larvae and next explaining the survival of larvae to the three year age class. Before making any regressions we note that spawning takes place outside of the Barents Sea, hence data pertaining to the Barents Sea will not be sufficient to explain development in the earliest phase of the herring's life.

Table 11 shows results for larvae. Since 1983 was a year with an extremely strong age class, the table shows results with and without this year. Linear models are used. Spawning stock of herring comes out with the expected signs. The model with data for 1983 data gets a R^2 of 0.41 while the one without gets 0.68. The latter model is also

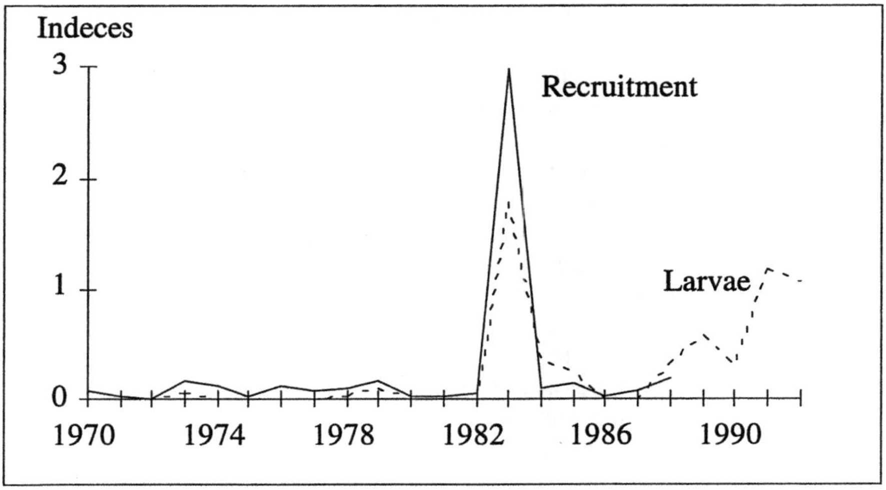

Figure 4. Larvae and recruitment of three year olds for herring. The VPA measure for three year olds is backdated to coincide with the corresponding larvae observations.

accepted at a much lower level of significance. Note however that the parameter estimates are not very stable, tests with additional explanatory variables typically makes the reported estimates change, even though the new variables produce no significant effects themselves. Prior information seems crucial to come up with an appropriate set of explanatory variables.

The test of the relationship between larvae and recruitment is shown in table 12. Again the results are shown with and without data for 1983. The interesting finding is that larvae explains much of the variation in recruitment. Hence, larvae measurements serve as an early indicator of recruitment. The coefficient is dependent on the data point for 1983. This might indicate a non-linear relationship or that 1983 is different from the other years by chance.

Young herring is found to be a significant predator on herring recruits. The coefficient varies in pace with the coefficient for herring larvae. We find no significant effect of cod, temperature and plankton.

Cod recruitment

As for herring, cod recruitment is measured in terms of larvae (a logarithmic measure of a half year olds taken in august or September) and in terms of a VPA measure for three year olds, see figure 5. Again spawning takes place outside of the Barents Sea, in areas for which we have not gathered data.

Table 13 shows that we are not able to come up with a good explanation of cod larvae. Temperature is likely to play a role, the sign is the same as for capelin and herring. The effect of mature cod comes out with the opposite sign of what one should expect from stock-recruitment models.

Can recruitment be explained by larvae and other variables? Table 14 shows the results. R^2 is 0.78 and the level of significance for the whole model is very low. Cod larvae comes out with the expected sign and is significant at the 0.5 percent level. Temperature is also significant. Young herring comes out as a significant predator, and capelin recruitment the following year is found to be a prey. Capelin and young cod are not found to have significant effects as predators. Herring recruitment the

Table 11. Estimates for herring larvae, 1973-1991 ($R^2 = 0.41$; N = 19; Whole model, Prob>F = 0.0015). Excluding observations for 1983: ($R^2 = 0.68$; N = 18; Whole model, Prob>F = 0.0002).

Term	Estimate	Std Error	t-Ratio	Prob>|t|
Intercept	-1.63	0.74	-2.21	0.042
Spawning stock	0.21	0.14	1.52	0.15
Temperature	0.35	0.16	2.28	0.037
Excluding observations for 1983:				
Intercept	-0.46	0.39	-1.19	0.25
Spawning stock	0.32	0.068	4.69	0.0003
Temperature	0.082	0.084	0.98	0.34

Table 12. Estimates for herring recruitment, 1973-1988 ($R^2 = 0.98$; N = 16; Whole model, Prob>F = 0.0000). Excluding observations for 1983: ($R^2 = 0.53$; N = 15; Whole model, Prob>F = 0.0107).

Term	Estimate	Std Error	t-Ratio	Prob>|t|
Intercept	0.08607	0.03643	2.36	0.0344
Herring larvae	1.6036999	0.0647	24.79	0.0000
Young herring	-0.162222	0.03469	-4.68	0.0004
Excluding observations for 1983:				
Intercept	0.0862496	0.01397	6.17	0.0000
Herring larvae	0.4806541	0.13091	3.67	0.0032
Young herring	-0.04507	0.01889	-2.39	0.0344

Table 13. Estimates for cod larvae, 1970-1991 ($R^2 = 0.18$; N = 22; Whole model, Prob>F = 0.15).

Term	Estimate	Std Error	t Ratio	Prob>|t|
Intercept	-1.15	1.33	-0.87	0.40
Mature cod	-0.19	0.19	-0.99	0.33
Temperature	0.48	0.26	1.81	0.086

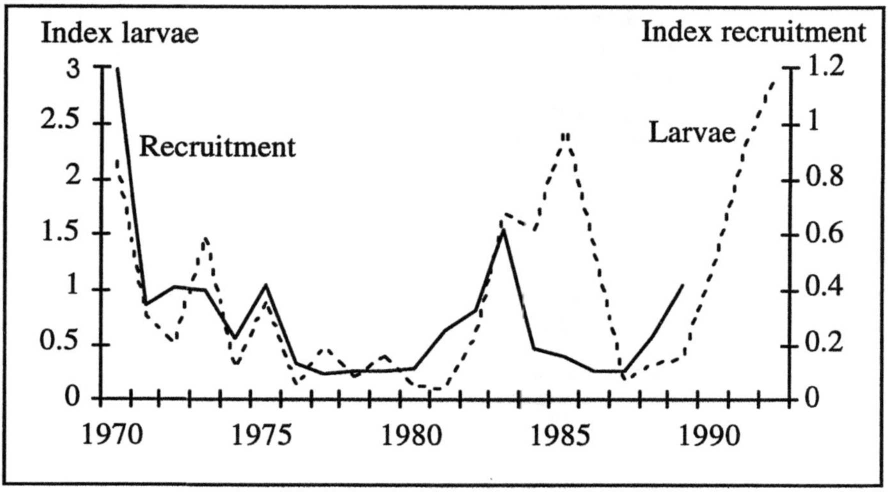

Figure 5. Indices for VPA cod recruitment (3 year olds), and larvae of cod, logarithmic and "linear" scale. Source ICES (1993): Cod table 3.14b, larvae table A13. Recruitment data are backdated to coincide with larvae data

following year is not found to be a significant prey.

Finally, note that while the larvae estimate comes out with the expected sign and as significant, it is not very important for predictions. When removed from the regression, the R^2 drops from 0.78 to 0.60. Temperature is more important since R^2 drops to 0.44 when this variable is excluded.

Temperature

An autoregressive model has been used to test for autocorrelation in temperature. Yearly September temperatures for the period 1921 to 1992 are used. A R^2 of only 0.18 is obtained for a model with the temperatures for the four most recent years. Only the one year lag seems to be fairly stable when different subperiods are used. The correlation is positive and the coefficient varies between 0.2 and 0.4. Between 5 and 15

Table 14. Estimates for cod recruitment, 1971-1989 ($R^2 = 0.78$; N = 19; Whole model, Prob>F = 0.0002).

Term	Estimate	Std Error	t-Ratio	Prob>|t|
Intercept	-0.65	0.18	-3.72	0.0023
Temperature	0.16	0.034	4.69	0.0003
Cod larvae	0.13	0.039	3.37	0.0045
Cap.recruitm. next year	0.06	0.023	2.92	0.011
Young Herring	-0.087	0.032	-2.76	0.015

percent of the variation is explained by the one year lag. Thus, an autoregressive model is not able to explain much of the temperature variation. The residuals for the one year lag model are accepted as normally distributed by a Shapiro-Wilk W test. The distribution seems to be truncated. We have not tested for the presence of cycles.

COMPARISON TO HISTORICAL DEVELOPMENT

According to our rough interpretation of Bayesian statistics, we like to keep the most certain prior estimates, and use the whole model's confrontation with time series data to calibrate the most uncertain parameters or to rethink model formulations. We indicate below that this is not a simple, straight forward task.

The most important random variable is recruitment. Using recruitment as an exogenous variable is problematic since recruitment to the zero class is unknown. Either we have to rely on larvae data, or we have to neglect behaviour in year classes below the point at which recruitment is normally reported. Here we report a test of the latter approach. An attempt to let the model predict recruitment from larvae observations was not very successful, see Moxnes (1994). Note that the simulations we perform are ballistic such that state variables are not corrected to take account of past observations as would be the case when using a Kalman filter.

Figure 6 shows model development together with historical observations from 1972 to 1992 for the case when behaviour in lower age classes is ignored. Recruitment, equal to observed recruitment, enters the second age class for capelin, the third for herring and the fourth for cod. In the model, the lowest age classes have been decoupled by setting the accompanying parameters for b, b_h, g, l, s, and m equal to zero. This choice of course unables us to study the interactions between the lowest age classes, it is only the appropriateness of the model for higher age classes that is tested. In addition, the exclusion of lower age classes implies that some of the preys for higher age class predators are removed. Hence, even if the model was absolutely correct for higher age classes, we should not expect the model to replicate history perfectly. Note also that lack of recruitment data for herring for 1989 and on, introduces a downwards bias for herring. Historical records of catch for each species and for temperature enters the model as exogenous variables. Temperature influences plankton only.

The major problem with the simulation is that the biomass of capelin comes out much lower than the observations during the seventies. It is difficult to point out a major reason for the deviation. Changing the initial conditions, by allowing for much more capelin in the first year class in 1972, only helps to restore the balance for a year or so. Capelin recruitment could be underestimated although the average capelin recruitment for the years from 1972 to 1976 was slightly higher than the average for the entire period. The model produces a low simulated plankton abundance, due to unfavourably high temperatures. We have no historical data for plankton abundance

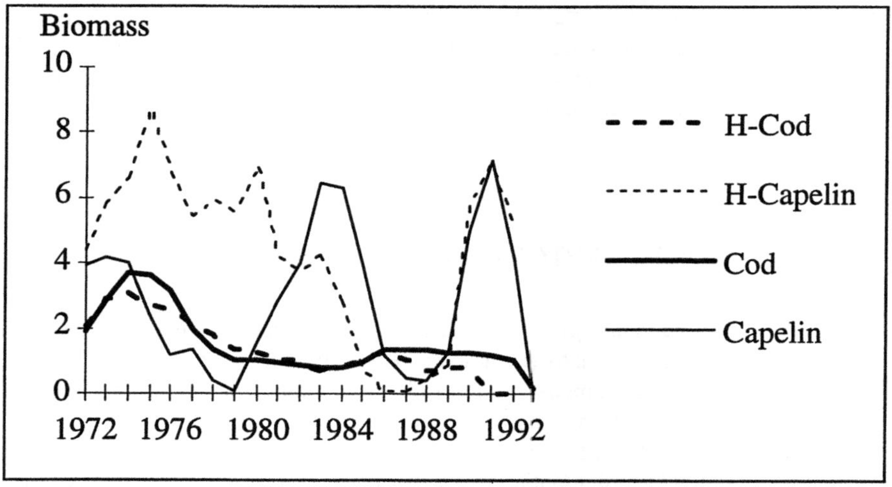

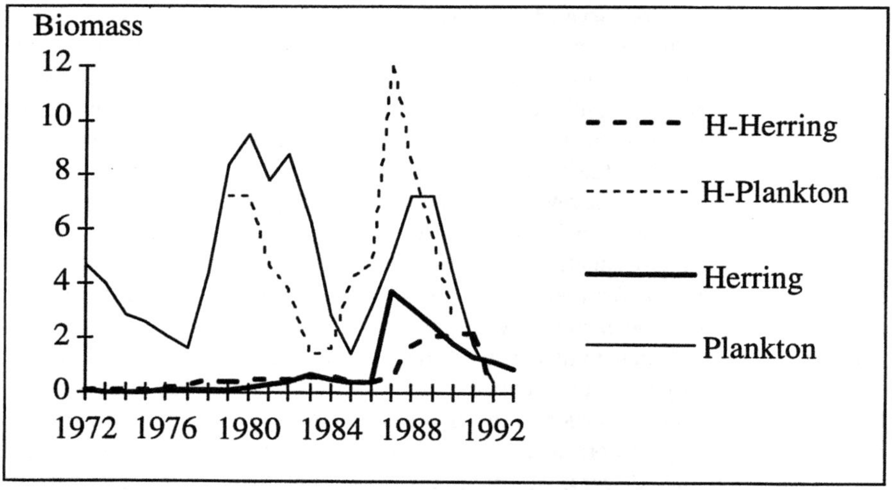

Figure 6. Simulated and historical (H-) biomass developments. Reported recruitment enters respective age classes. Lower age classes are ignored. Plankton is divided by 7.5 to fit the scale.

for most of the seventies to judge this potential source of error. In a test, we find that an exogenous reduction in the temperature variable by 0.5°C for the period 1972 to 1975 serves to maintain plankton around its initial level, and the capelin biomass becomes higher. However, capelin still tends towards zero at the end of the seventies, at least partly explained by higher cod biomass. In the simulation, predation loss for capelin reaches a maximum for the seventies of 2.5 million tonnes per year, in 1974. The simulated cod population at this point in time was slightly higher than the historical observations. Simulated development for herring is lower than historical records from 1973 to 1984. A common explanation for capelin and herring seems likely. Rather systematic measurement errors is also a possible explanation.

From 1980 to 1983 the conditions are reversed, simulated plankton is higher than historical records and cod biomass is equal to or lower. Both capelin and herring experience stronger growth than in the historical data. After 1983 all species behave fairly correctly. Herring shows deviating behaviour at the end of the period due to lacking recruitment data. It also seems that the 1983 year class enters the biomass data more abruptly in the simulation than in the data. Plankton is roughly in line with the existing observations.

CONCLUSION

We have explored the possibilities for constructing an aggregate model of the multispecies fish resources of the Barents Sea, useful for the purpose of policy analysis under uncertainty. A simplified version of existing multispecies models, Multspec and Aggmult, was constructed and discussed in light of existing literature. An effort was made to enable direct use of prior information in the model.

In section 3 several interactions were explored statistically by the use of time series data. In general this exercise was encouraging since in many cases coefficients came out with expected signs, significance levels for both parameters and models were low, and at times the models did explain quite large fractions of the variation. These results were obtained in spite of measurement errors and short time series. The presence of autocorrelation in many of the time series might have contributed somewhat to the apparently good results.

We found that the results do not contradict the chosen formulation of predation. An aggregate model for plankton, explains most of the variation in plankton density in terms of temperature variations. Recruitment for capelin is explained very well in terms of predators, plankton and temperature. These variables turn out to be much more important than the biomass of capelin, the variable usually used in stock-recruitment relationships. Larvae production for herring seems to be explained by the biomass of herring to a fairly large extent. Herring recruitment is fairly well explained by herring larvae. Only predation on herring larvae from young herring is found to be significant. Cod larvae are not well explained, not even in terms of cod biomass. Cod recruitment on the other hand seems to be explained by the supply of larvae, the abundance of capelin larvae, predation from young herring, and by temperature. Temperature is not well explained by its own history.

A simulation shows that the model is quite successful in explaining historical development of biomass for the different species when records of recruitment are used as exogenous variables and lower age classes are excluded from the model. The major exception is capelin and herring in the seventies. Another simulation shows that the model is not successful in explaining recruitment from larvae data.

In spite of its weaknesses, the model and the regressions seem to be a good starting point for further model improvements and for stochastic policy analysis. The model for the higher age classes is at a stage where it can be used to test policies, given specifications of probabilities for error terms. However we have experienced that the model is complicated and time consuming to work with in many respects. Therefore it is interesting to search for model simplifications. Such simplifications should make it easier to calibrate and work with the model.

Finally a few words about how the model, or simplified versions of it, can be used for policy analysis. The "curse of dimensionality" when using dynamic programming implies that truly optimal policies under uncertainty cannot be found for large models. However, it is computationally feasible to find optimal linear approximations to policy functions for rather large models. Monte Carlo simulations are then used to find expected criterion values for different parameter combinations in a presumedpolicy function. A nonlinear search procedure is used to find the parameter set that maximises the expected criterion value. See e.g. Walters (1986). The loss caused by deviating from the truly optimal policies through simple approximality seems small in many cases.

Logical restrictions on the policies such as requiring non-negativity is a simple and straight forward extension into non-linear policy functions. Further complications can be added from knowledge about truly optimal policies for simplified models. Fine tuning such non-linear policies to more complicated models, provides an improvement of current knowledge even though truly optimal policies are not obtained. Further improvements can be obtained by using intuition and by allowing more flexible functional forms.

It seems wise to start the policy analysis using simple models and than add complexity step by step. This approach has been taken in Moxnes (1996) where a two species model is analysed. Quota policies for cod and capelin are found as functions of the biomass for the two species. First the above method is shown to yield results comparable to those obtained by standard stochastic dynamic progamming for the simplest version of the model. Then the method is used to show how policies change when non-linearities are introduced in the functions for predation, costs, and prices, and when the criterion is extended to value stability.

The same method can be used to analyse the value of precision in measurements of resources, and to estimate the value of precision in model parameters, see Moxnes (1997) for applications to a two species model. When analysing the value of better measurements, optimal policies are found for different degrees of precision. The marginal values of precision follows from the corresponding changes in expected criterion values. When analysing the value of precision in parameter estimates, optimal policies are again found for different degrees of precision, and marginal values of precision in parameter estimates are calculated. If marginal costs of precision in measurements and in parameter estimates are known, optimal accuracy's can be found.

REFERENCES

Andersen, K.P. & E. Ursin 1977. A multispecies extension to the Beverton and Holt theory of fishing with accounts of phosphorus circulation and primary production. Medd. Danm. Fisk- og havunders. 7: 319-435.

Bogstad, B., S. Tjelmeland, T. Tjelta & Ø. Ulltang 1992. Description of a multispecies model of the Barents Sea (MULTSPEC) and a study of its sensitivity to assumptions on food preferences and stock sizes of minke whales and harp seals. Draft, SC/44/O 9, Institute of Marine Research, Bergen.

Eide, A. 1991. ECONMULT 3.0. Manual, University of Tromsø, Tromsø.

Eide, A. 1993. Fluktuasjoner i torskebestanden. Working paper, University of Tromsø, Tromsø.

Forrester, J. & P. Senge 1980. Tests for building confidence in system dynamics models. TIMS Stud. Man. Sci. 14: 209-28.

Hatlebakk, E., G. Hagen, K. Merckoll & J. Hamre 1993. System modell for havfiske. Note, STAT/05/1993, Norwegian Computing Center, Oslo.

Holling, C.S. 1965. The functional reponse of predators to prey density and its role in mimicry and population regulations. Mem. Ent. Soc. Can. 45: 1-60.

ICES 1992. Report of the Atlando-Scandian herring and capelin working group. Copenhagen.

ICES 1993. Report of the Artic fisheries working group. Copenhagen.

James, M.K. & K.P. Stark 1982. Application of the Three Bays Ecosystem Model to Fisheries Management. In: Pauly, D. & G.I. Murphy (eds.). Theory and management of tropical fisheries. ICLARM Conference Proceedings 9. Manila: International Center for Living Aquatic Resources Management.

Larkin, P.A. & W. Gazey 1982. Application of Ecological Simulation Models to Management of Tropical Multispecies Fisheries. In Pauly, D. & Murphy, G.I., ed. Theory and management of tropical fisheries. ICLARM Conference Proceedings 9. Manila: International Center for Living Aquatic Resources Management.

Moxnes, E. 1994. Stockfish, a Multispecies Model for Stochastic Analysis. SNF-Report 82/94, SNF, Bergen.

Moxnes, E. 1996. Optimal quota policies for a sthochastic predator -prey system. SNF -report 76/96, SNF, Bergen.

Moxnes, E. 1997. Two-species Management under uncertainty: The value of accuracy in measurements and model parameters. Report (preliminary), SNF, Bergen.

Skjoldal, R., H. Gjøsæter & H. Loeng 1992. The Barents Sea ecosystem in the 1980s: ocean climate, plankton, and capelin growth. In ICES mur. Sci. Symp. 195.

Sparre, P. & R. Willmann 1993. BEAM IV, a bio-economic multi-species, multi-fleet, multi-plant, multi-area extension of the traditional forecast model. In: Nielsen, J.R. & T. Helgason (eds.). Multispecies Fisheries Management. Proceedings of a symposium in Hirtshals May 22-23 1992. Copenhagen: Nordic Council of Ministers.

Spulber, D.F. 1985. The Multicohort Fishery Under Uncertainty. Mar. Reso. Econ. 1(3): 265-282.

Walters, C.J. 1986. Adaptive Management of Renewable Resources. New York: Macmillian.

Zellner, A. 1981. Philosopy and Objectives of Econometrics. In: Currie, D. et al. (eds.). Macroeconomic Analysis: Essays in Macroeconomics and Econometrics. England: Croom Helm.

Bioeconomic Multispecies Models of the Barents Sea Fisheries

By Arne Eide and Ola Flaaten.
The Norwegian College of Fishery Science, University of Tromsø, Norway.

INTRODUCTION

It is common knowledge among marine biologists that big fish eat smaller fish and that the smaller fish eat even smaller fish, crustaceans and plankton. Theoretical biologists started studying ecological systems by means of mathematical models many years ago (e.g. Lotka,1925 and Volterra,1928), and studies of single species models are even older (e.g. Verhulst,1838). The breakthrough in the use of mathematical models in applied marine-biological research came with the Baranov-Beverton-Holt (BBH) model and the Schaefer-model in the 1950s. The BBH model is a single species model in which the relationship to other species is not explicitly formulated. However, in a way these relations are implicitly taken care of. The constant mortality rate includes mortality due to predation by other species; the growth rate and recruitment to the fishable stock implicitly depend on availability of prey and on the presence of competitors.

Single species models of this kind have many advantages compared to more complex ecosystem models: they are quite simple, the parameters have a specific biological meaning and they can in many cases be estimated from catch, effort and research data. Most important, however, is that for a long time the single species models proved to be sufficiently reliable for management purposes. This is especially the case when the harvesting has targeted just one, or a few species in an ecosystem, and when the level of fishing effort directed against different species has been changing slowly with time.

The technical and economic developments have led to commercial exploitation of more and more fish stocks, and stocks that sustained fisheries for a very long time have been severely depleted. Examples of the latter type of fisheries are numerous in the North Atlantic, both in Canadian and Icelandic waters, as well as in the North Sea and the Barents Sea. This is probably the main reason for the increased interest which biologists and others have taken in the use of multispecies models in applied research and management.

The main purpose of this paper is to report on some ongoing bioeconomic multispecies modelling work at the Norwegian College of Fishery Science, University of Tromsø (NCFS). The presentation will be limited to models applied to the fish stocks and the fisheries of the Barents Sea and adjacent areas. The main focus is on the multifleet model ECONMULT (Eide, 1991; Eide and Flaaten, 1994) and the use of this in combination with the biological multispecies model AGGMULT (Tjelmeland and Bogstad, this book). The latter is a multispecies cohort model currently comprising the three fish species cod (*Gadus morhua*), capelin (*Mallotus villosus*) and herring (*Clupea harengus*). AGGMULT is developed at the Institute of Marine Research, Bergen (IMR), and the development of this and ECONMULT is done in close cooperation between the researchers involved in the work of the two institutions. The work took place within the framework of, and was partly funded by, the multispecies program of the Norwegian Research Council.

In the mid 1980s the IMR started a systematic research on multispecies interactions in the ecological system of the Barents Sea and adjacent area. The aim was to make this work an integrated part of IMR's research and to design and implement a detailed multispecies model for this area. The model, named MULTSPEC, is documented in Tjelmeland and Bogstad (1989). Even long before this particular multispecies research started, Norwegian, Russian and other scientists have investigated different parts of the Barents Sea ecological system (for references, see Flaaten, 1988, as well as other papers in this book). The MULTSPEC model is very detailed and includes i.a. length groups and sex of fish, and their distribution among areas of the sea. For bioeconomic analysis it is more convenient to use the AGGMULT model.

The next section of this paper gives a short description of the ecosystem of the Barents Sea."Introduction to the economics of multispecies harvesting" gives an introduction to multispecies bioeconomic analysis to highlight the difference between this and biological multispecies analysis, and between singlespecies and multispecies bioeconomic analysis. "ECONMULT -an economic multispecies model" describes the ECONMULT model and the use of it in management scenarios. "ECONMULT and AGGMULT simulations" gives the ECONMULT simulation results for the period 1996-2025."Partial multispecies bioeconomic analysis of other species" presents results from other bioeconomic multispecies analysis of the Barents Sea fisheries, in particular for sea mammal - fish and for cod - shrimp interactions. The last section concludes the paper.

THE ECOSYSTEM AND THE FISHERIES OF THE BARENTS SEA

The Barents Sea occupies 1.405 million square km between the coast of North Norway, Svalbard, Novaja Semlja and the Murman Coast. The south-western part of the Barents Sea never freezes to ice, and the northern and eastern parts are ice-free for part of the year.

The total number of zooplankton species of the Barents Sea is relatively few, consisting of 145 species (Zenkevitch, 1963). The copepods make up the main zooplankton biomass: 90% in the south-western part of the sea. As in other seas of the world, the copepods are the basic food components for fish and other inhabitants. In addition to the zooplankton, there is another important main source of food for the fish: the benthos inhabiting the sea bottom.

There are 114 known species of fish inhabiting the Barents Sea, of which the commercially most important families are: the cod family (12 species) and the flounder (11 species). There are especially two species of zooplankton preying fish which have been of great commercial value: capelin (*Mallotus villosus*) and herring (*Clupea harengus*). The pelagic fish species are important food sources for larger fish species, sea mammals and birds. So also is krill (*Euphauciacea*), feeding primarily at the second trophic level. In the transference of energy from phytoplankton through the food web to fish, krill is an important link (Mauchline and Fisher, 1969).

Larger fish of high commercial value are cod (*Gadus morhua*), haddock (*Melanogrammus aeglefinus*) and saithe (*Pollachius virens*). On average for the years 1950-1990, these three species yielded approximately 90% of the total annual catch of demersal and semipelagic species (excluding pelagic species like herring, capelin and polar cod). Other demersal species of commercial value are: red fish (*Sebastes sp.*), halibut (*Hippoglossus hippoglossus*), Greenland halibut (*Reinhardtius hippoglossoides*), plaice (*Pleuronectes platessa*). Of high-valued species, deep water prawn (*Pandalus borealis*) and salmon (*Salmo salar*) are the most important ones. However, the commercial importance of the latter one has dwindeled due to the growth of aquacultured salmon.

The main food sources of cod are known consisting of small pelagic fishes such as herring, capelin, young cod, haddock and polar cod and large planktonic crustaceans. (see e.g. Zenkevitch, 1963; Mehl, 1989).

The 7 species of seals of the Barents Sea and adjacent areas are great consumers of fish, krill, zooplankton, benthos, etc. The most important species for commercial purposes has been harp seal (*Pagophilus groenlandicus*). Its opportunistic feeding behaviour (see e.g. Nilssen, 1995) is probably a very important explanation of the large stock to be found in the Barents Sea, including the White Sea.

Fish and seals occupy the Barents Sea area all year round, whereas 17-18 species of whales utilise the area through intensive feeding in summer. Since the early 1930s the minke whale (*Balaenoptera acutorostrata*) has been the most important commercially exploited whale species. The minke whale is an opportunistic feeder, consuming several species of fish, krill and zooplankton (see e.g. Haug et al., 1995).

For the Barents Sea fisheries cod, herring, capelin and deep water prawn have been among the commercially most important species. During the last fifty years it has been a significant decline in the total biomass of the cod population in the Barents Sea. The lowest stock levels were reached during the late 1980s. Hard regulations were introduced, especially in 1990 when the total allowable catch (TAC) reached a historical minimum of 160 thousand tonnes of North-east Arctic cod, i.e. the Barents Sea cod.

The average catches and biomass of cod and capelin are shown in table 1 for some periods of time. Note the significant decline in the average annual catch, as well as in the average stock level of cod.

The population growth of cod is characterised by large variations between year

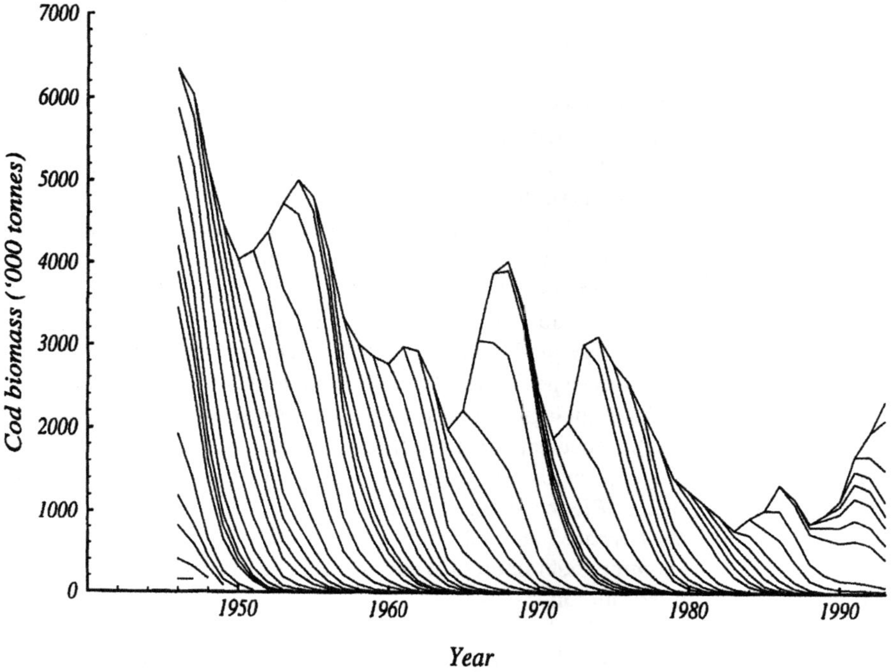

Figure 1. Biomass of cohorts and stock (3+) of Arcto-Norwegian cod 1946-93. Source: Based on data from ICES CM 1994/ ASSESS:2.

Table 1. Average annual catches and stock size of cod and capelin of the Barents Sea area 1953-92 (in '000 tonnes).

Period	Average annual catches		Average biomass	
	Cod[1)]	Capelin[2,3)]	Cod (3+)[4)]	Capelin[3)]
1953-62	863.0	n.a.	3655	n.a.
1963-72	717.4	659.4	2755	n.a.
1973-82	677.9	1870.9	2012	4520
1983-92	343.1	680.2	1178	1920
1963-92	579.5	1070.2	1982	-

Sources: [1)] ICES CM 1985/ Assess 2; Institute of Marine Research's Fisken og Havet, Særnummer 1, 1994.
[2)] ICES Bul. Stat. Vol. 48-65;
[3)] Institute of Marine Research's Fisken og Havet, Særnummer 1, 1987 and 1994.
[4)] ICES CM 1994/ ASSESS:2

classes. Even in years with a large cod stock, a few yearclasses tend to dominate the stock in biomass terms. Figure 1 shows the biomass of the Barents Sea cod stock from 1946 to 1993, as the cumulative biomass of following cohorts (year classes). Each year a new yearclass reach the age of 3 year and by that is recruited to what is considered as the fishable cod stock. The biomass of the new yearclass can be observed in the figure as a new top layer. When time goes the yearclass will grow older and the biomass of this yearclass will reach a maximum before it declines and vanish. The line that describes the growth of this cohort will be followed by new cohorts and new top layers. The envelope curve then describes the total biomass of all cohorts (3+) at any time. The vertical distance between the end point of the two uppermost curves for 1946 represent the biomass of 3 year old cod this year, i.e. the 1943 yearclass.

Even though Barents Sea cod may become 20-30 years old, the International Council for Exploration of the Seas (ICES) usually lump together those 15 year and above. The oldest ones are not noticeable in figure 1. This is because few fish reaches the highest ages. Figure 1 shows, e.g., that the cod stock consisted of 12 and 6 noticeable yearclasses in 1946 and 1965, respectively.

INTRODUCTION TO THE ECONOMICS OF MULTISPECIES HARVESTING

This section gives a brief description of the difference between biological and economic management criteria, and the difference between single and multispecies analysis. The implications for management are also indicated. Some elements and results from a three species model of the Barents sea fisheries (Flaaten,1988) will be used to illustrate our points.

In a singlespecies biomass model the sustainable yield (SY) curve will usually be a concave function of the stock size. The maximum point on the curve represent the maximum sustainable yield (MSY) of the fish stock, and this is often seen as a reference point for biological management. From an economic point of view, however, other points on the SY-curve could be optimal, depending on the price of fish and cost of effort involved. If the price of fish, p, is a non increasing function of the harvest, h, and the cost per unit harvest, c, is a non increasing function of the stock size, X, the net value per unit harvest, b, is

$$b(h,X) = p(h) - c(X) \tag{1}$$

From an economic point of view management could mean, firstly, to derive the equilibrium stock size that maximises the total net value, $P=b\cdot h$, called the resource rent. Secondly, to find and use policy means to control fishing effort and/or harvest. However, even if this may sound easy to do, it rarely is in real management cases. This book, and several others, are proofs of the difficulties involved. Still, we want to highlight some similarities and differences between single- and multispecies management approaches, as well as between biological and economic management criteria. Therefore, the analysis is kept as simple as possible.

In addition to the economic weighting of the physical harvest by the net value per unit harvest, b, it is usually necessary to assign weights to income and costs at different points in time to make them comparable. This problem of discounting is solved by giving relative lower weights to future revenues and costs than to present. The discount factor assigned to period t is

$$d_t = e^{-\delta t} \quad (2)$$

The recommended value of the discount rate, δ, used in cost-benefit analysis of Norwegian public investment projects, is 7 per cent p.a. This is in real terms, i.e. adjusted for inflation. The use of a positive discount rate has implications for the long term optimal stock level, as well as for the timepath towards the targeted long run level. Biological management criteria are lacking the types of weights given in (1) and (2). Biological knowledge of recruitment, individual growth and natural mortality is in most cases a necessary condition for sound economic management of a single fishstock, but it is not sufficient.

Going from singlespecies to multispecies models and analysis usually implies making one or more of the parameters, e.g. recruitment, growth and mortality, into endogenous variables. This type of research belongs to the biologists and other natural scientists. From an economics research point of view the ideal situation would be to have the complete ecological model with all parameter values known, either with "certainty" or with mean values and variance. The economic research would then concentrate on the studies of demand-, harvest- and cost functions entering into the relative weights given by (1) for the different species involved.

A singlespecies biomass model gives the sustainable yield curve, the MSY rate and the corresponding stock levels, whereas the cohort model gives a yield per recruit plot as a function of the first age of harvest and the fishing mortality. As a parallel to the former model the multispecies biomass model would give a n-dimensional yield surface in the n-dimensional room. In the two species case this implies that the maximum yield of one species is a function of the yield of the other. This combination of yields is called the maximum sustainable yield frontier (MSF) and is the long run limit to what can be harvested from the two species. It is, however, important to note that harvesting less than a combination of yields on the MSF does not necessarily secure a sustainable harvest. For example, in the case of a predator-prey model biological overfishing of the prey may contradict the underlying assumptions for the MSF. In this respect two species MSF harvesting is to be compared to MSY harvesting in singlespecies models.

The main problem for the manager of the ecosystem is to choose the best, in one or another meaning, combination of yields for the two species. With a positive discount rate and the objective of maximising the present value of future rent from the resource, the long run strategy will be to aim at a combination of yields below the MSF curve. This is due to the capital costs of the resource stocks.

To illustrate the biological limits to harvesting in the case of three interacting

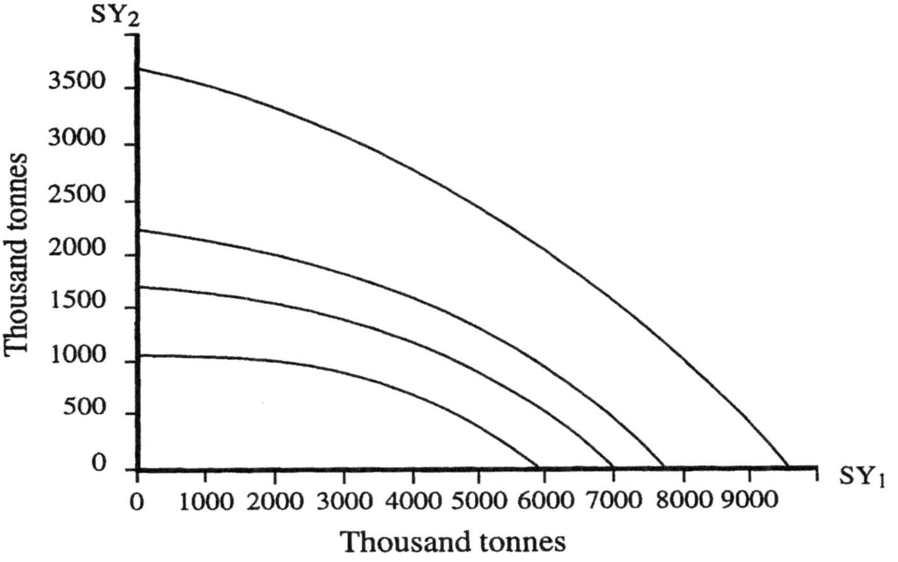

Figure 2. Maximum sustainable yield frontiers (MSF) for species one and two for four different stock levels of species three. (From Flaaten, 1988.)

species, an alternative to the three dimensional yield surface is the conditional two dimensional MSF curves shown in figure 2 (for details, see Flaaten, 1988). Species one is the basic prey eaten by species two and three, and species three also feed upon species two. Thus, in this case species three is the top predator of the ecological system. The condition for each of the MSF curves in figure 2 is a given stock level of species three. The innermost MSF curve is for the highest stock level of species three, whereas the outermost curve is the theoretical boundary for MSF of species one and two if species three becomes extinct.

Assume that the preferences for the harvest of the three species are given by such weights as in (1), and the relative weight of future to present income, the discount factor, is given by (2). Then maximisation of the present value of rent from the harvesting of the three species can point out the optimal long run stock levels and sustainable yields to aim at. However, in some cases there could be utilities affiliated to the proper stock level of one or more species. For example, if species three is a sea mammal, some people prefer a high stock level to a low one. The manager may now maximise the present value of the future rent, given the restriction that species three is equal to or higher than a given stock level. This means that one of the MSF curves in figure 1 is the limit to the long run yields of species one and two. For disaggregated multispecies models, for example models with several species, ageclasses and length groups, it is usually not possible to derive analytical optimal solutions. In such cases computer simulations can be useful. That is the approach used in the following section of this paper.

ECONMULT - AN ECONOMIC MULTISPECIES MODEL

The original goal with the ECONMULT modelling project was to develop a tool that could be useful for the fisheries management administration, particularly in the management of the Barents Sea fisheries. Soon it became clear that from a management point of view it would be necessary to have several models, highly aggregated models and more disaggregated models, depending on the problem to be focused. Therefore, it was necessary to develop several models, each focusing on different problems. This called upon a more flexible solution, a solution that could be used on a broader range than on management problems only.

Another important reason for establishing a more flexible framework, was to make it possible not only to update parameters as new data become available, but also incorporate improvement in the more basic knowledge about the dynamics of the system. All within the original framework of the model.

The history of the model

The modelling project was initiated by the authors, at the NCFS, and partly financed by the Norwegian Research Council, as a part of the Multispecies Management Program (MMP) in 1991.

As a framework ECONMULT represents the core unit of several bioeconomic models developed within the MMP. The first model to be presented was the combined ECONMULT/MULTSIMP model, also named ECONSIMP (Eide, 1991; Eide and Flaaten, 1994). MULTSIMP is a biological model developed by the IMR (Tjelmeland, 1990). The original core unit of ECONMULT has remained more or less unchanged since it first was used in the ECONSIMP model. The parameters of the fleet model have been updated several times. The most significant change was the introduction of the AGGMULT model in 1992. AGGMULT has the same structure as ECONMULT (see Tjelmeland, 1992). This means that the flexibility of the fleet model now has become available also in the biological model.

The first ECONMULT/AGGMULT model included the cod and capelin fisheries of the Barents Sea (Eide, 1992). Today AGGMULT also include herring, which soon will be included in the fleet model.

A short description of the model

ECONMULT is a framework of a fleet model and represents a tool for creating fisheries models on different aggregation levels.

The basic idea of the ECONMULT concept is:

1. To establish a simple interface between a multispecies biological cohort model and an economic fleet model.
2. In a simple way to make it possible to use the same model concept while studying problems that needs different resolutions at different levels.

ECONMULT itself does not contain any biological module, but the use of it require a compatible biological model to be connected to the ECONMULT fleet model. Several models have been designed within the framework of ECONMULT. Up to now the focus has been on management problems, but in principle ECONMULT could be used also in developing models of more scientific or commercial interest.

The variables of the ECONMULT model

There exist two types of variables in ECONMULT: Structure variables and Model variables. The Structure variables define the resolution of the specific fleet model, while the Model variables represent ordinary variables in the fleet model that has been defined.

There are three types of Structure variables in ECONMULT:

1. Variables that define the structure of the fleet model: Number of vessel groups, Number of targeted species, Number of separate biomass units (cohorts).
2. Bolean switches for turning on or off different features of the model.
3. The time unit.

Normally the model variables are matrices. The size of each matrix is determined by the setting of the type 1 Structure variables. Model variables can be separated into:

1. Decision variables.
2. Endogenous variables.

What is decision variables and what is endogenous variables is partly determined by what type of management that regulate the fisheries. In an open access fishery no decision variables will be active, while a directly regulated system could have different decision variables being active at different times. Variables that always remain endogenous are the biomass and the number of fish in each cohort of the stocks defined in the model.

Model outline

ECONMULT is a simulation model with one or several groups of homogenous fishing vessels. Each group may harvest one or more of the fish stocks, and each stock may include one or more cohorts. A vector of cohort biomasses is an important interface between the biological and the economic part of the model. Altogether there are s stocks and u cohorts, as shown in figure 3. A fishery is defined as a combination of one vesselgroup and its main targeted fish stock. However, every cohort may in principle be caught in every fishery, either as being part of the main targeted stock or as bycatch. With m vessel groups there are altogether $m \cdot s$ fisheries in the model, and $m \cdot s \cdot u$ types of catches. Future extensions of the model may include geographical distribution of the fish stocks.

Figure 4 shows a general outline of the ECONMULT-model. For a specific objective functional the fisheries should be managed to maximise this objective. The harvest rates for the cohorts in one period will be deducted from the cohort biomasses before the start of next period. The cohort biomasses and catches thus are very important links between the biological and the economic part of the multispecies model.

In the presentation of the mathematical equations of the model the following symbols are used.

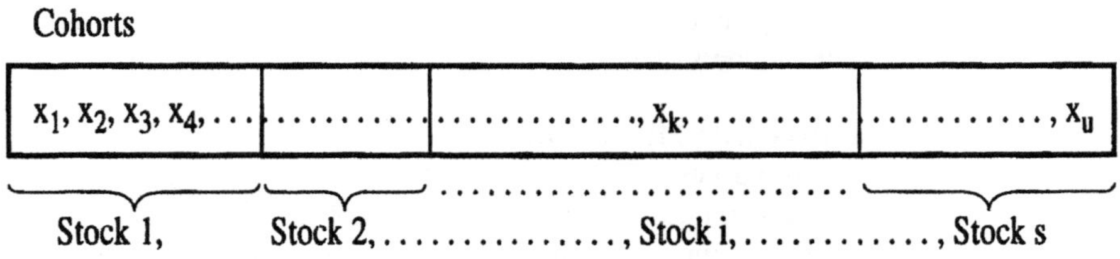

Figure 3. The relationship between the ECONMULT-concept of stock and cohorts. The number of cohorts in each stock may vary.

Subscripts:

i Refers to stock
j Refers to vesselgroup
k Refers to cohort

List of variables and parameters:

a_{ijk} Pricefunction's constant term
b_{ijk} Pricefunction's variable term constant
cm_{ij} Contribution margin
d_{ijk} Cohort's area distribution and availability for catch
e_{ij} Days of fishing per vessel (fishing effort per vessel)
fc_{ij} Fixed costs
h_{ijk} Vesselcatch of one cohort
h_{ij} Vesselcatch of all cohorts in a given stock
q_{ijk} Catchability parameter
m Number of vesselgroups
n_{ij} Number of vessels in each fishery group
p_{ijk} Ex-vessel price of fish
s Number of stocks
tc_{ij} Total vessel costs
tr_{ij} Total vessel revenues
u Number of cohorts in the model
vc_{ij} Variable vessel costs (costs per day of fishing)
x_{ik} Biomass of one cohort
C Priceaffecting catch
E_{ij} Total fishing effort of the vesselgroup
H_{ijk} Vesselgroup's catch of given cohort
H_k Total catch of cohort
X_{ijk} Biomass of given cohort available in the fishing area of the vesselgroup
a_{ijk} Effort - output elasticity
b_{ijk} Stock- output elasticity
δ Discount rate
p_{ij} Vesselprofit
P_i Vesselgroup profit

All the variables are related to specific periods of time. Some parameters are constants, other may vary between periods. To simplify the notation, subscripts are omitted in the equations that follow.

The catch of a specific cohort in a given fishery is modelled by the Cobb-Douglas harvest function:

$$H = q \cdot E^{\alpha} \cdot x^{\beta} \tag{3}$$

here $E = n\ e$. Catch per vessel is

$$h = \frac{H}{n} = q \cdot e^{\alpha} \cdot n^{\alpha-1} \cdot x^{\beta} \tag{4}$$

Ex-vessel price of fish is either constant or a linear function of a specific segment of the catch:

$$p = a - b \cdot C \tag{5}$$

where C includes relevant catch of one or more cohorts from the vessels in one or more fisheries. The pricefunction may be on stock basis or on cohort basis.

The revenues of the vessels come from the sale of fish:

$$tr = p \cdot h \tag{6}$$

The harvesting costs consist of fixed costs, which is mainly capital costs, and variable costs, which are proportionate to the number of fishing days.

$$tc = fc + vc \cdot e \tag{7}$$

This means that the average total cost of fishing effort is decreasing with increasing effort per vessel.

With constant ex-vessel prices of fish the contribution margin per vessel in a given fishery is

$$cm = p \cdot h - vc \cdot e \tag{8}$$

and the profit (included resource rent) is

$$\pi = cm - fc \tag{9}$$

In case of elastic demand the constant price in equation (6) should be substituted by the price function (5).

The total profit (included resource rent) of a given fishery is

$$\prod = n \cdot \pi \tag{10}$$

and the present value of the profit/rent is

$$PV = e^{-\delta t} \prod(t) dt \tag{11}$$

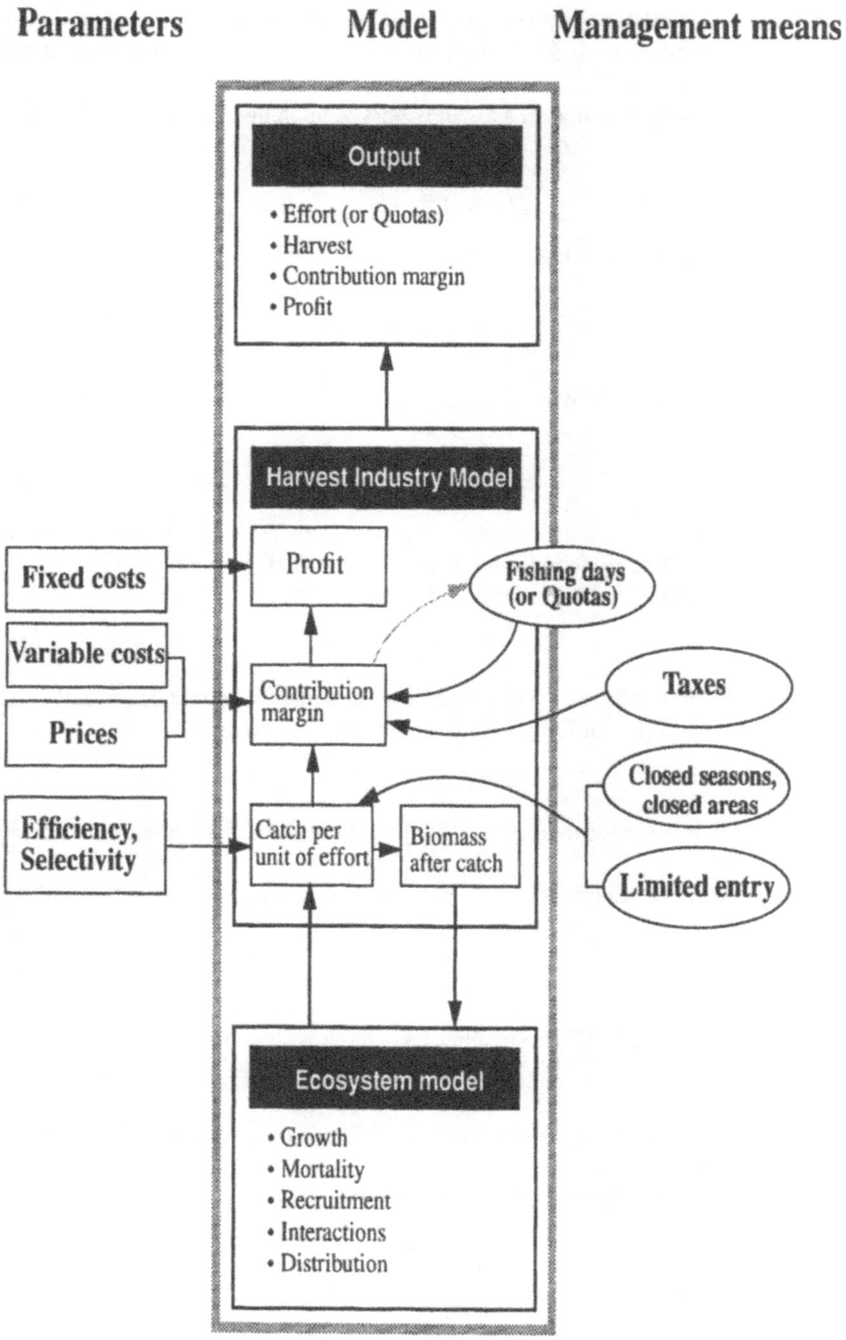

Figure 4. The structure of the ECONMULT model.

Using the model

The fisheries of ECONMULT may be simulated by using one of the following combinations of management means:

i) fishing effort per vessel (e) and catch quota per vessel (h),
ii) fishing effort per vessel (e) and total catch quota (H),
iii) number of vessels (n) and fishing effort per vessel (e),
iv) number of vessels (n) and catch per vessel (h),
v) number of vessels (n) and total catch (H).

In the short run the number of vessels are fixed and the procedures i) or ii) may be the most relevant. In the long run, however, the number of vessels are not fixed, and the procedures iii) - v) may be more useful. The details of the simulation procedures are explained in Eide (1992). The simulations presented in the next section are mainly based on procedure v).

ECONMULT AND AGGMULT SIMULATIONS

According to the mutual agreements between Norway and Russia the Barents Sea cod stock is shared equally between them, and the capelin is shared 60-40 percent in the favour of Norway. These distributions of the total allowable catches (TACs) of cod and capelin are used in all ten simulations. In this analysis we assume that the two parties have mutually interests and apply the same means. However, the reported economic figures are always that of Norway's. We disregard the fact that other countries than Norway and Russia in recent years have got about 10 percent of the TAC of cod.

Previous results

Before presenting new results by use of ECONMULT and AGGMULT, previous results obtained by using a more simple biological model MULTSIMP; Tjelmeland 1990 will be referred.

Nine different simulation runs from the joint model of MULTSIMP and ECONMULT has been published by Eide and Flaaten (1994). Starting with stocklevels estimated by IMR for January 1, 1989, using reported catches for 1989, 90 and 91, the proper simulation period commenced with estimated model stocks January 1, 1992 and runs for another 27 years. The number of vessels in each vesselgroup in the constant effort period were related to 1989 or 1990 figures for the cod fisheries and to 1991 for the capelin fisheries, due to the 1987-90 moratorium for the latter. The number of vessels were held constant during the 27 year simulation period, and a constant 7% p.a. social rate of discount was used.

Of the nine simulations the best economic result was obtained by a constant fishing effort 50% higher than the 1990- effort for the cod fisheries and 50% higher than the 1991-effort for the capelin fisheries. For the Barents Sea fisheries 1990 was a very bad year with no capelin quota and a record low quota of cod. It is obvious from the simulation results that interior solutions with simultaneous fishing of cod and capelin

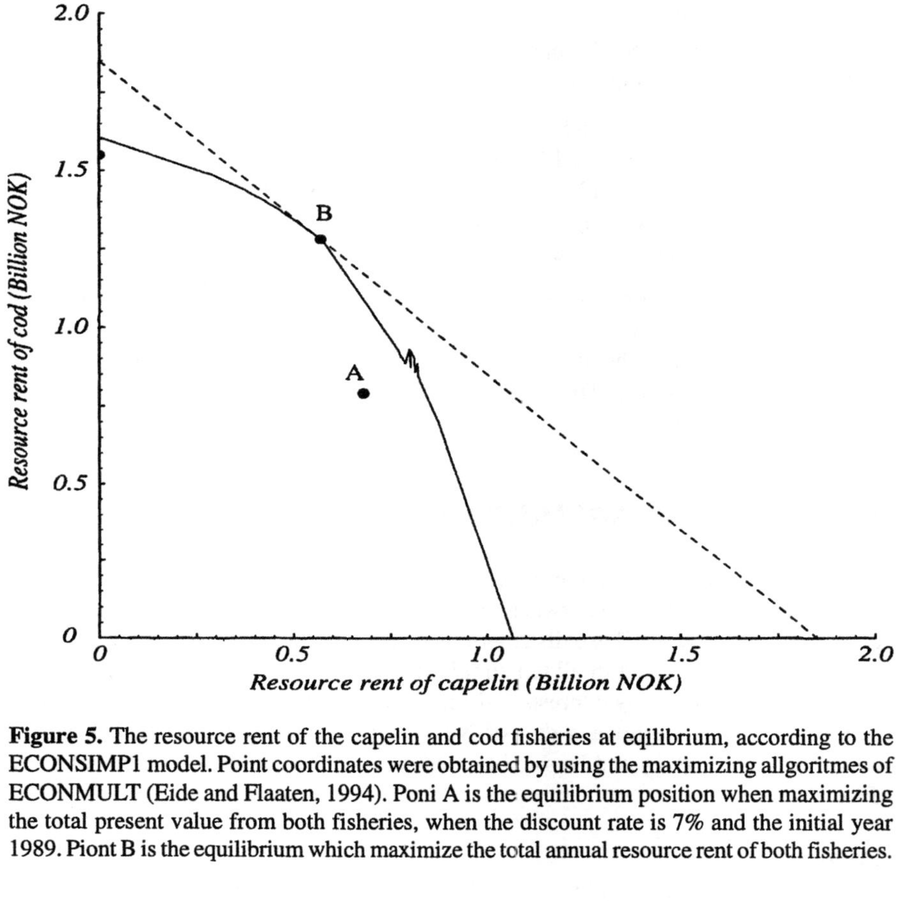

Figure 5. The resource rent of the capelin and cod fisheries at eqilibrium, according to the ECONSIMP1 model. Point coordinates were obtained by using the maximizing allgoritmes of ECONMULT (Eide and Flaaten, 1994). Poni A is the equilibrium position when maximizing the total present value from both fisheries, when the discount rate is 7% and the initial year 1989. Piont B is the equilibrium which maximize the total annual resource rent of both fisheries.

are better than no harvesting of either of the two stocks. The worst result appear when there was no fishing of the cod stock, and the second worst with no capelin harvest.

Using only the bioeconomic most cost-effective vesselgroups within the cod and the capelin fisheries, respectively, maximising sustainable economic rent gave the total rent of 1.85 billion Nkr. per year, of which 70% came from the cod fishery. Increasing the price of cod 25%, ceteris paribus, increased both the total economic rent and the cod fishery's share. Increasing the price of capelin, ceteris paribus, also increased the total profit compared to the basic case. In this case the optimal effort level in the capelin fishery increased by 20% and the effort level in the cod fishery increased by 8%.

In all simulations the opportunity cost of labour equals the average hourly earnings of workers in the Norwegian manufacturing industries. Data on average annual days of fishing and working hours per day at sea are obtained from mainly the Directorate of Fisheries' annual costs and earnings studies.

The resource rent equilibrium curve for the cod fisheries and the capelin fisheries is shown in Figure 5. The optimal solution that maximises present value of rent is, according to Eide and Flaaten (1994), below the curve at point A, in a position of combined harvest of cod and capelin.

ECONMULT and AGGMULT simulations

The simulations presented in this section are based on the latest version of AGGMULT (see Tjelmeland and Bogstad, this book). The estimated values of the parameters of AGGMULT are based on data from the period 1973-92. Except for the recruitment parameters, all parameters are fixed for the whole period of simulation. The variations in recruitment parameters reflect changes in the environment, e.g. sea temperature, currents and plankton production. Thus, the simulations for the 30 year period 1996-2025 are based on the assumption that the underlying biological conditions from 1973-92 will be repeated in a cyclic way (see Tjelmeland and Bogstad, this book).

By varying exogenous variables, e.g. total allowable catch (TAC) or fishing mortality (F), the effects on the biological variables, such as total biomass and yearclass biomass, and on economic variables such as contribution margin and resource rent, may be studied. When fishing take place in the model, there must be one or more restrictions on the catches to avoid depletion or extinction of the stocks. Effective restrictions may be TAC, number of vessels or lower limits on the contribution margin or resource rent. In Norway there is limited entry into most fisheries and in many cases this is combined with the use of national TACs and other quota arrangements.

Types of simulation

The differences among the simulations are as follow:

Type A (simulations 1-5 and 7).
- -Contribution margin is positive in all quarters (cm>0).
- -Limits on the annual catch of cod (TAC_{cod} is restricted).
- -No restrictions on fishing mortality of cod (F_{cod} varies over years).
- -No capelin fishery ($F_{cap}=0$).

Type B (simulation 6).
- -Capelin fisheries, winter and summer ($F_{cap} \geq 0$)
  Otherwise as for type A.

Type C (simulation 9).
- -Labour costs are included in the variable costs.
  Otherwise as for type A.

Type D (simulation 8).
- -Contribution margin may be negative in some quarters (cm < 0).
  Otherwise as for type A.

Type E (simulation 10).
- -Fishing mortality of cod is restricted ($F_{cod} \leq F_{cod}$).

-No fixed limit on the annual catch of cod (TAC_{cod} varies over years). Otherwise as for type A.

Data and reference periods

The simulations start with the actual biomass of stocks and yearclasses as of January 1 1994. For all 10 simulations the actual TACs for 1994 and 1995 are restrictions on the fisheries for these two years. For each of the 10 simulations shown in table 2 the management means are kept constant for 30 years from January 1 1996.

This version of ECONMULT has 29 vessel groups, but no. 1 and no. 2 are the same vessels participating in the summer and winter fishery for capelin, respectively. The number of vessels in each vessel group and quarter is shown in the appendix, table A1. The effort-output elasticities are, by assumption, equal to one for all vessels. The stock-output elasticities vary somewhat between vessel groups and are shown in the appendix, table A1. The catchability coefficients vary between vessel groups, yearclasses and quarters, and are shown in the appendix, table A3.

Price and cost data are from 1989, but are adjusted for inflation to reflect the price level January 1 1996 (see the appendix, table A1 and A2).

Results

The results of 10 simulations are shown in table 2. As noted above the TACs are divided between Norway and Russia, whereas the reported economic figures are that of Norway only. Of type A simulations, where annual TACs of cod are the main regulatory means, simulation 3 with TAC equal to 375 thousand tonnes of cod, gives the highest present value of contribution margin. This also gives the highest average catch for the 30 year simulation period. For TACs up to 500 thousand tonnes the quota is an effective limit for the annual catch in one or more years. The TAC of 750 thousand tonnes is, however, not an effective constraint on the catch in any of the years 1996-2025.

With TAC regulations the average biomass of cod clearly decreases with an increase in the quota, from 2,495 thousand tonnes for TAC equal to 300 thousand tonnes in simulation 1, to 1,156 thousand tonnes for TAC equal to 750 thousand tonnes in simulation 7. The annual variations in catches of cod are greater the higher the TAC is. The biomass of cod towards the end of the 30 years period decreases with an increase in the TAC. For TAC equal to 300 thousand tonnes the biomass of cod January 1 2025 is more than three times as high as when the TAC equals 750 thousand tonnes. The biomass of mature capelin at the end of the 30 year simulation period shows a quite clear increase with an increase in the TAC for cod.

Biomass and harvest during the 30 years period are shown in figure 6, panel a and b, respectively, for four TAC simulations. Simulation 1 with the lowest TAC clearly has a long term increase in the stock. In this case the quota effectively becomes binding in most years. The annual catches are very close to each other most years in the cases of TAC equal to 500 and 750 thousand tonnes. This is because the quota hardly becomes binding in any year in these two cases. Only in a few of the peak years does simulation 7 (TAC=750 thousand tonnes) give higher annual catch than

Table 2. Simulation results.

Type of simulation					Results of simulations (30 years, 1996-2025)							
	Management means			Present value[1] (mill. NOK)	Catch of cod ('000 tonnes)			Cod biomass (age 3+, '000 tonnes)				Capelin biomass (Mature stock, '000 tonnes)
No.	TAC_{cod} '000 tonnes	F_{cod}	Notes		Average(st.dev.)	Min.	Max.	Average(st.dev.)	Min.	Max.	Jan.1 2025	Jan.1 2025
1	300	Free		13,450	289.7 (23.0)	213.8	300.0	2495.1 (650.2)	1448.8	3768.5	3768.5	728.6
2	350	Free		14,774	323.9 (54.5)	147.1	350.0	2098.9 (563.1)	741.6	3047.9	2787.0	1194.6
3	375	Free		14,932	333.6 (67.4)	133.7	375.0	1788.4 (486.3)	722.3	2634.6	2001.8	1656.2
4	400	Free		14,380	319.0 (98.5)	110.7	400.0	1511.4 (502.8)	608.4	2283.5	1657.8	1685.4
5	500	Free		13,381	297.7 (137.6)	84.3	500.0	1189.6 (501.8)	337.9	2243.2	1214.8	1994.8
6	500	Free	2	13,960	297.6 (137.6)	84.3	500.0	1188.7 (502.0)	337.0	2243.2	1212.6	1980.3
7	750	Free		13,232	294.8 (144.8)	72.5	562.8	1156.4 (507.7)	313.7	2262.3	1216.0	1970.1
8	750	Free	3	7,239	265.5 (175.2)	60.0	652.2	714.0 (505.0)	127.0	1807.2	589.5	2256.2
9	750	Free	4	19,435	328.1 (122.9)	112.7	543.4	1964.4 (439.2)	1244.5	2675.6	2122.9	1217.4
10	Vary	≤ 0.3		14,025	314.6 (131.4)	107.5	548.0	1332.2 (471.7)	577.0	2246.1	1345.7	1758.5

1. Present value January 1, 1996 of contribution margin of Norwegian catch, i.e. revenues net of operation costs, labour costs excluded. A discounting rate of 5% has been used.
2. Summer and winter capelin fishery, limited entry but no catch quota.
3. Possible with negative contribution margin, full capacity during all seasons.
4. Labour costs included in the contribution margin evaluation . (No fishing while the contribution margin become negative.)

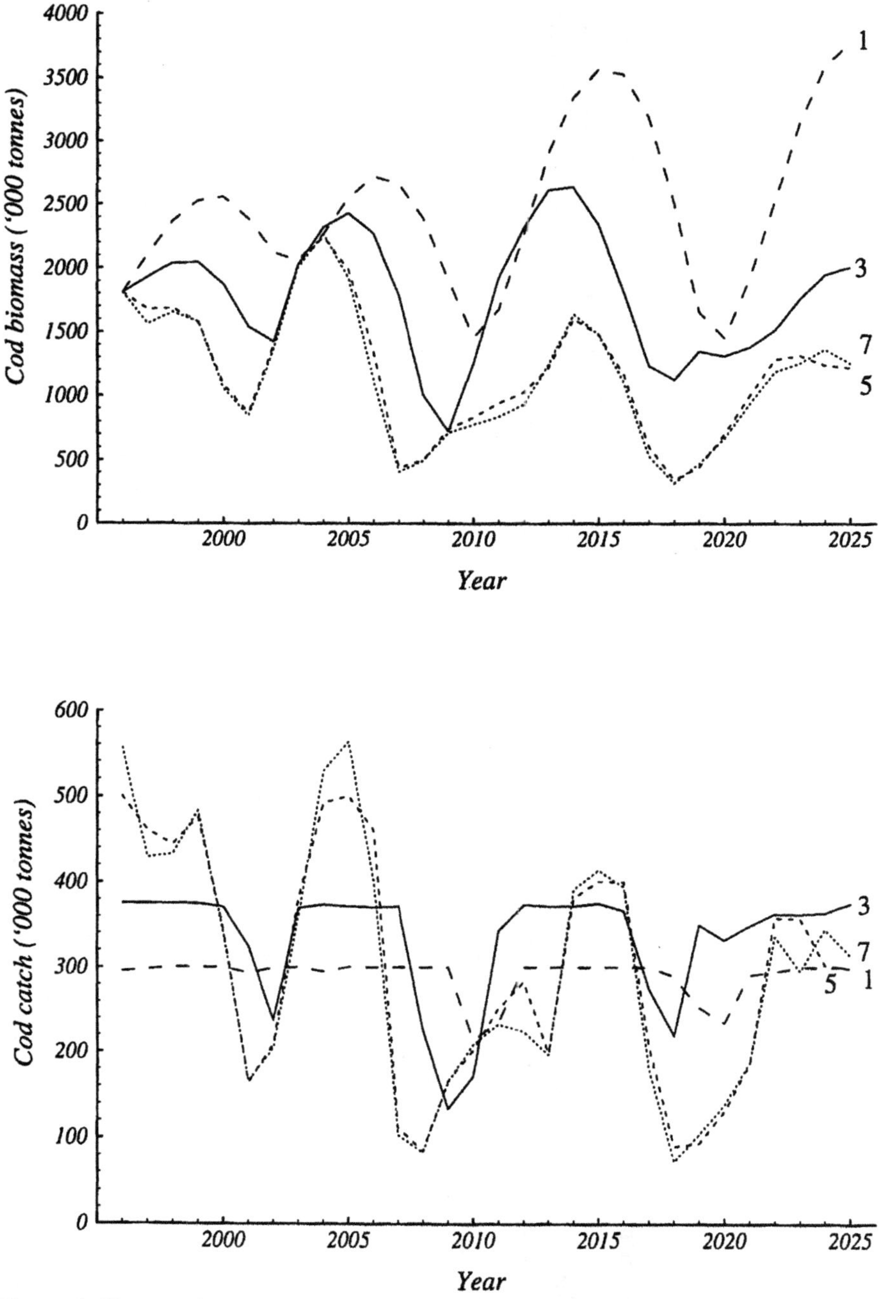

Figure 6. The simulation results for the Arcto-Norwegian cod stock's total biomass and annual catches, for TACs of 300, 375, 500 and 750 (simulations 1,3,5 and 7) thousand tonnes, 1996-2025.

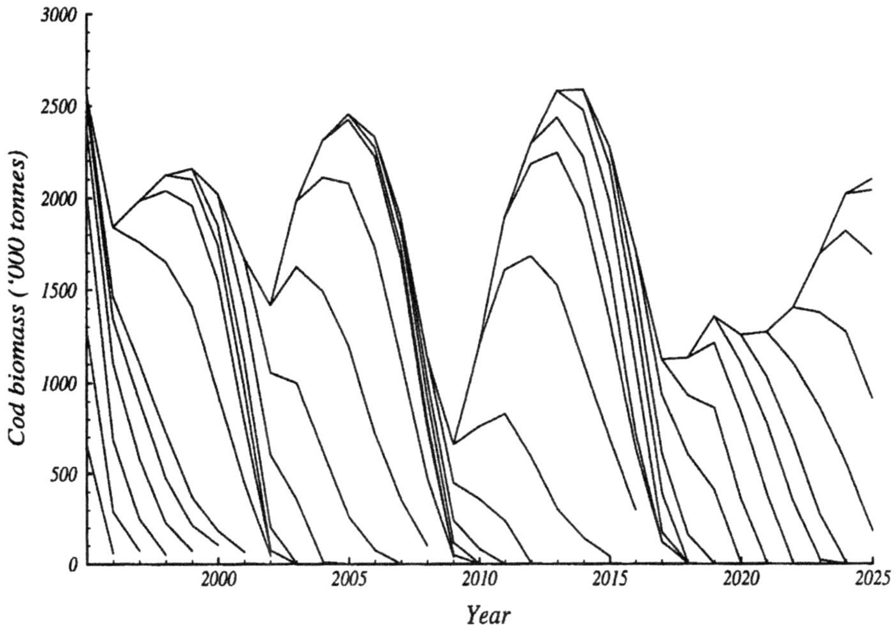

Figure 7. The simulation results from the Arcto-Norwegian cod stock's total biomass and distribution among yearclasses for an annual TAC of 375 thousand tonnes (simulation 3), 1996-2025.

simulation 5 (TAC=500 thousand tonnes). The result of this is that the development of the stock is very close to each other in simulations 5 and 7, as shown in figure 6 panel a.

Figure 7 shows the biomass of the cohorts and the cod stock during the 30 year period 1996-2025 for simulation 3. Comparing this to the historic development 1946-1993 in figure 1 there is one striking difference in the pattern. In the historic figures the stock seem to consist of more age groups than in the simulated future. On the other hand there is a similarity between the figures in that there are strong variations in year class strength and in total biomass in both cases.

Simulation 6 allows for winter and summer capelin fishery, whenever there is a positive contribution margin. This is the only simulation with capelin fishery. The results, shown in table 2, include some increase in the present value of contribution margin in the combined cod and capelin fisheries. However, for all biological criteria shown in table 2, there is very little effects on the harvest and the stock of cod from the capelin fisheries. During the simulation period the capelin stock biomass is at a very low level. Even in the case of an open access capelin fishery (simulation 6) the capelin stock size is larger at the end of the period than earlier in the period, as is the case of all the other simulations. The capelin biomass values of table 2 therefore also represent the maximum capelin biomass during the whole period for all simulations.

For simulation 9 labour costs are added to the variable costs shown in the appendix

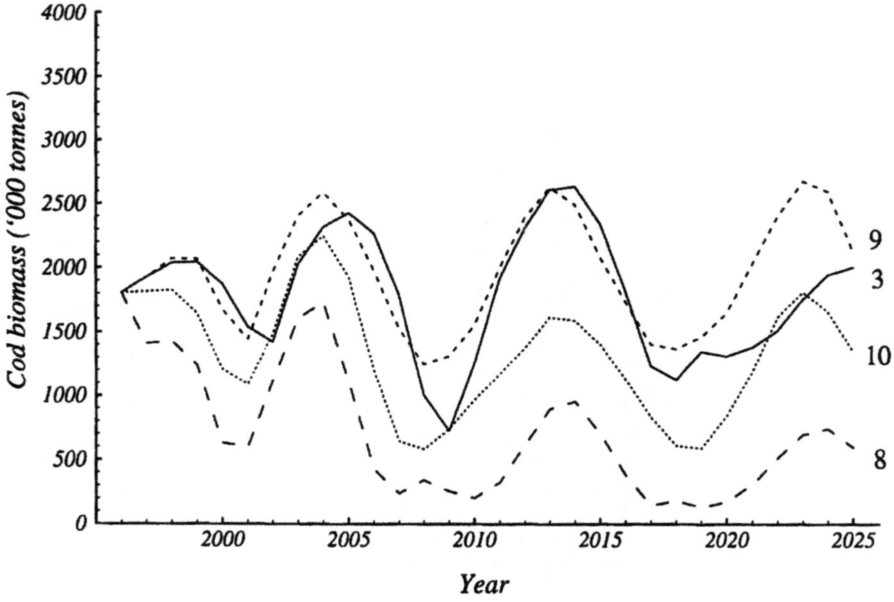

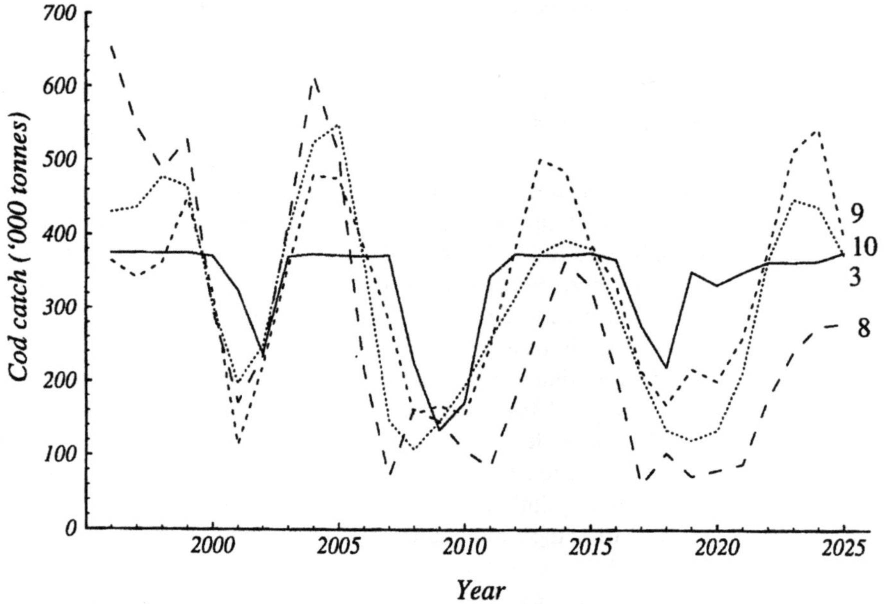

Figure 8. The results for the Arcto-Norwegian cod's total biomass, panel (a), and annual catches, panel (b), for simulations 3, 8, 9 and 10, 1996-2025.

table A1. (The average hourly earnings of workers in manufacturing industries affiliated with the Confederation of Norwegian Business and Industry was in 1989 NOK 87.25 according to Norwegian official statistics.) This means that fishing will take place only in those quarters where the revenues are greater than the sum of variable costs and labour costs, provided that the TAC does not get binding. This results in a significant higher minimum stock level of cod than for the other simulations, except for no. 1. The maximum stock level, however, is approximately the same as in simulation 3. The present value of the contribution margin for the 30-year period is significantly higher for simulation 9 than for all the other simulations. Note that the definition of the contribution margin in equation (8) and the variable costs are the same for all 10 simulations.

Simulation 8 allows for full capacity utilisation in any quarter even if the contribution margin becomes negative, provided the TAC does not become binding. As shown in table 2 this results in the lowest present value of contribution margin among the 10 simulations. This simulation also has the lowest average and minimum stock levels. Comparing simulations 7 and 9, economic restrictions on fishing require that in the latter case the stock level of cod has to be higher than in the former case. Whereas comparing simulations 7 and 8, economic restrictions are looser in the latter case allowing fishing at lower stock level than in the former one.

Simulation 10 has a restriction on the annual fishing mortality, i.e. $F \leq 0.3$. This value is between F_{med} and F_{max} used by The International Council for Exploration of the Seas (see ICES, 1993). Given $F=0.3$, the annual TAC will vary with the stock size. Table 2 shows that the average annual catch in simulation 10 is almost the same as for simulation 4, with TAC equal to 400 thousand tonnes, but with greater fluctuations in the annual catches. Simulation 10 also has greater fluctuations in the stock size, and the stock level in year 2025 is lower than for simulation 4. The economic result of simulation 10 is, however, almost as good as for simulation 4.

Figure 8 shows the simulated biomass and catch of cod for simulations 3, 8, 9 and 10. Recall that simulation 3 is the best one from an economic point of view. Simulation 3 has the least variations in the annual catches, whereas simulation 8 has the greatest difference between the minimum and the maximum catches. Negative contribution margin are allowed in simulation 8 and this causes the stock being fished down to its lowest level in any of the 10 simulations.

With the exception of simulations 8 and 9, there are relatively small differences in present value among the other simulations. However, there are considerable differences in stock levels of cod and capelin, valued either by the average over the years 1996-2025, or especially, by the level the last year.

Discussion of the simulations

In the ten simulations three different types of constraints on the cod fisheries have been investigated. The constraint represented by a constant TAC during all years, will lead to a low fishing mortality (F) when the stock level is high, and a high fishing mortality on low stock levels. On the other hand, the constraint on the contribution margin will, when active, reduce the fishing mortality when the stock level falls below the critical level of a positive contribution margin. The third constraint (or management mean) which has been investigated in simulation 10, is a constant fishing mortality. In addition to these three constraints, the fixed number of vessels represent a limited entry regulation in all ten simulations. This regulation becomes active when the limited number of vessels do not have the capacity to harvest the TAC during a year, while the contribution margin is positive. Without any limitation on entry to the fisheries, the number of vessels in that case should be expected to increase.

Six different constant annual TAC values have been tried out, simulating a period of 30 years of fishing (1996-2025). They are all of type A (simulations 1-5 and 7). The TAC values are chosen on basis of the historical catch of cod during the last 30 years (Table 1). One of these simulations (TAC = 750 thousand tonnes) does not make the TAC restriction active, due to the additional constraint of the contribution margin, which has to be positive to make the vessels fish. This economic constraint becomes active some years in all simulations, also at the lowest TAC level of 300 thousand tonnes.

Even though a low TAC value gives a more constant catch over all years (Figure 6), the fluctuation of the cod stock will be substantial. This is partly due to the environmental fluctuations which is built into the recruitment process described by AGGMULT, but also due to an increase in cannibalism when the cod stock grows in biomass. Simulations 5 and 7 follow almost the same catch path (See Figure 6). The annual catch in simulation 7 exceeds 500 thousand tonnes, which is the TAC of simulation 5, only a few times. Following almost the same catch path, simulation 5 and 7 are very similar also in all other terms, such as the present value of the contribution margin (Table 2). Following a very different catch path, simulation 1 still is almost similar to simulations 5 and 7 in present value terms, whereas a TAC value of 375 thousand tonnes in simulation 3 gives a significant increase in the present value. The difference between the present value of the latter simulation and simulations 1, 5 and 7 is about 1.5 billion NOK (Table 2). If the stock level is kept lower than in simulation 1, the cannibalism in the cod stock is reduced, while the cost effect of a more severe depletion of the cod stock (as in simulation 5 and 7) is avoided in simulation 3.

Figure 7 shows the fluctuations and age composition of the cod stock in simulation 3. Strong yearclasses in 1999 and 2000 together with unfavourable environmental conditions leads to a very weak recruitment from 2002 through the following four years. This is the reason for the low biomass level in year 2009, about 720 thousand tonnes. At the end of the 30 year period the fluctuations seem to be reduced. A parallel increase in the capelin biomass may reduce the tendency of cannibalism during this period.

The huge fluctuations generated by the high constant TAC values, may involve costs not included in the model, such as lower ex-vessel prices when the catch increases rapidly from one year to another. On the other hand, a large cod biomass probably also represents costs not included in this model. Some of these predation costs will be discussed below.

Simulation 6 involves capelin fishery without quota limitations, only restricted by limited entry and the constraints represented by a positive contribution margin. Immature capelin is caught during the summer, while mature capelin is caught during the winter period. The capelin stock was almost extinct at the start of the simulation period, and does not recover before around year 2010. Even after this time the capelin fishery is at a very low level compared to the last 30 year history of the fishery (Table 1). The maximum catch during the simulation period was just above 675 thousand tonnes, but most years there were no catch at all. Including these years the average annual catch during the 30 year period is about 180 thousand tonnes. The difference between the present value of simulation 5 and 6 of about 600 mill. NOK (Table 2) is the contribution from the capelin fishery. Since most of the capelin catch is from the later part of the 30 year simulation period, the present value of this catch is very low compared to the present value of the cod fisheries.

According to this model one should not expect the capelin fishery of the Barents Sea to become of great importance during the period 1996-2025. Even in the cases of a reduced predation pressure by cod and no capelin fishery, as in simulation 5, 7 and 8, the mature capelin stock at the end of the period, will only be about 2 mill. tonnes. The strong effect of the predation pressure by cod on the capelin stock is demonstrated by the very low capelin stock size of about 730 thousand tonnes at the end of the period in simulation 1. Even in this case the size of the capelin stock has a maximum at the end of the period. In all simulations the capelin stock collapses four times during the period of 30 years. The break downs of the stock seems to be due to mainly poor recruitment.

The strongest reduction of the cod biomass is represented by simulation 8, where the constraint regarding the contribution margin is removed. In this simulation the fishing effort is constant over the year, and equal to the one of quarter 1 (Table A1) in the other simulations. Even when removing this constraint, the TAC value of 750 thousand tonnes is not reached in any year. The maximum catch is just above 650 thousand tonnes. As being the worst case, this management clearly shows the importance of the economic restrictions represented by the contribution margin of the vessel. The difference between simulation 7 and 8 indirectly shows the effect of subsidies, simulation 8 representing a fully subsidised fleet, although the total capacity of the fleet is not changed. In reality subsidies also should be expected to influence on the total capacity, which could give even worse scenarios than simulation 8.

Simulation 8 could represent a decrease in the variable costs (e.g. by subsidies), while simulation 9 shows the effect on the fishing activity when the variable costs increase. In simulation 9 the contribution margin will turn negative on biomass levels far above what was the case in the other simulations. The gain of reducing fishing effort earlier when the stock is declining, shows up in later periods where the catch is increased as a result of growth in biomass. This effect is easy to read out for Figure 7, where the line of simulation 9 gives the two last peaks of catch during the period. The significant difference between this simulation and all others in present value terms (Table 2) is from the increase in cost efficiency made by the fact that simulation 9 gives almost the same average catch as simulation 2 and 3, but using a different minimum level of biomass. The costs in catch units therefore has to be lower, while the revenue almost will be the same. Simulation 9 could also be looked at as a case of taxation. By introducing a tax on effort equal to the labour costs used, the results theoretically should be the same.

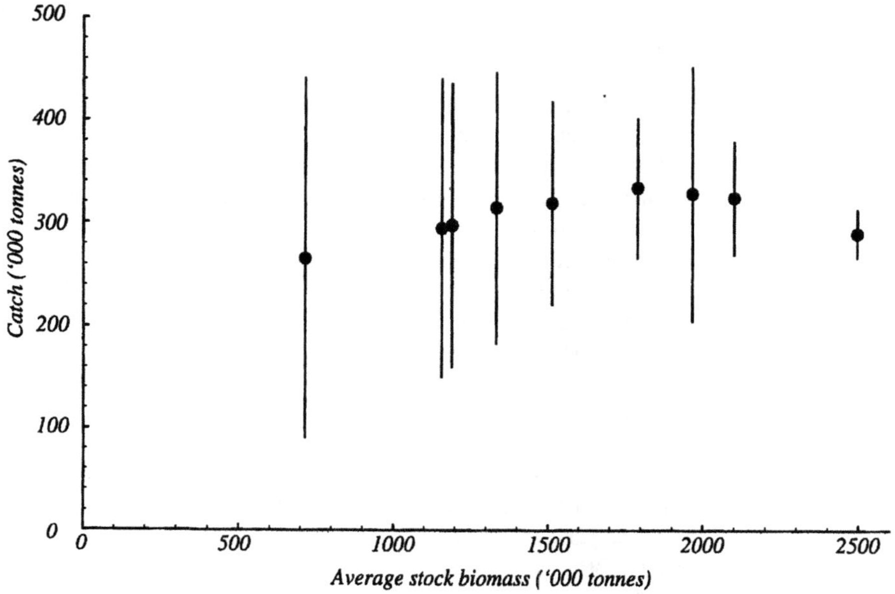

Figure 9. Average cod catch on average cod biomass of the simulations presented in Table 2. The numbers represent the simulation number (1-10). The vertical lines give the standard deviation of each simulation.

The last type of simulation (type E) is represented by simulation 10, where the TAC of each year is calculated on the basis of a management rule saying that the fishing mortality should equal 0.3 on an annual basis. The management rule is close to the type of management we have today. This fishing mortality covers all the stock of three year and older cod. This management rule implies a lower exploitation rate than a constant TAC does, on low stock levels. The effect of this is similar to the one of simulation 9, but weaker. Compared to simulations 2, 3 and 4 it is not obvious that a management rule of a constant fishing mortality is preferable to a constant TAC over a long period of time (Table 2 and Figure 7).

As shown in Figure 9 the standard deviation of the annual cod catch of each simulation (also see Table 2), tends to decrease as the average stock biomass increase. This indicate a more fluctuating catch quantity on lower stock levels. The exception is simulation 9, where the fishing activity is reduced more rapidly as the profitability goes down. This lead to a faster growth of the stock, which in the next periods causes larger catches. These changes in catches therefore increase the total profitability of the fisheries. However, one has to add that in these simulations the price is constant and not affected by the changes in catch from year to year.

PARTIAL MULTISPECIES BIOECONOMIC ANALYSIS OF OTHER SPECIES

This section review bioeconomic analysis of some species that have not been, or still are not included in AGGMULT. The AGGMULT model used in the ECONMULT simulations above now includes the fish species cod, capelin and herring (see Tjelmeland and Bogstad, this book). However, at the time when the research for this paper was undertaken, herring was not implemented in the biological model. In addition to the three fish species the MULTSPEC model now comprises minke whale and harp seal. It still remains to include these sea mammals in the AGGMULT model. Simulations with the MULTSPEC model shows that minke whales and harp seals may have a significant effect on the fish stocks in the Barents Sea and adjacent waters (see Bogstad et al., 1995).

Neither AGGMULT nor MULTSPEC at present include shrimp, which is a commercially important high valued species. However, as shown in figure 1 in Tjelmeland and Bogstad, this book, the biomass of shrimp annually consumed by cod in the Barents Sea may be of significant magnitude. There are, at least, two good explanations why shrimp is not the most obvious prey species of cod to include in a multispecies biological model. First, the population dynamics of Barents Sea shrimp is yet not very well understood. Second, the growth effect on cod from shrimp as prey is probably of less importance than the energy richer capelin and herring.

As an alternative to complete multispecies models, that include the total population dynamics of two or more species, it is possible to use simpler models that include only some of the ecological elements found in complete models. This section reports on such bioeconomic analysis of minke whale - fish and cod - shrimp interactions. Let us start with the analysis of cod as a predator, to see how the feed composition vary with age and how this could affect the economics of the prey fisheries.

Figure 10 shows the Northeast Atlantic cod's age-dependent average annual consumption of some commercially important prey species.

Species included are shrimp, capelin, herring and cod (cannibalism) above 5, 10, 10 and 20 cm, respectively. The figures are in grams prey per kg cod, for each age class of cod from 1 to 7+ years. Figure 10 shows, for example, that one kg of two year old cod annually consumed 2000 gram prey of these four species above the given size, and that about 75% of this was capelin. For all age classes, capelin is the main prey among the species and size groups included in figure 10.

Taking the net opportunity cost of feed into consideration (see Flaaten and Kolsvik, 1995 for details) gives the results shown in figure 11. The net value of the prey is the net contribution which the fish in the sea could have given for the prey harvesters if they had less competition from the predator, the cod.

The net value per unit catch was found by Flaaten and Kolsvik, 1995 to be 30% of the quay- side price in these fisheries. Figure 11 shows, for example, that two year old cod had an annual feed cost of NOK 1.50 (ECU 0.19) per kg of biomass, and that about 75% of this was inflicted on the shrimp fisheries. Except for age class 7+, the opportunity cost of shrimp dominates the economic figures, whereas capelin dominated the biological results in figure 10.

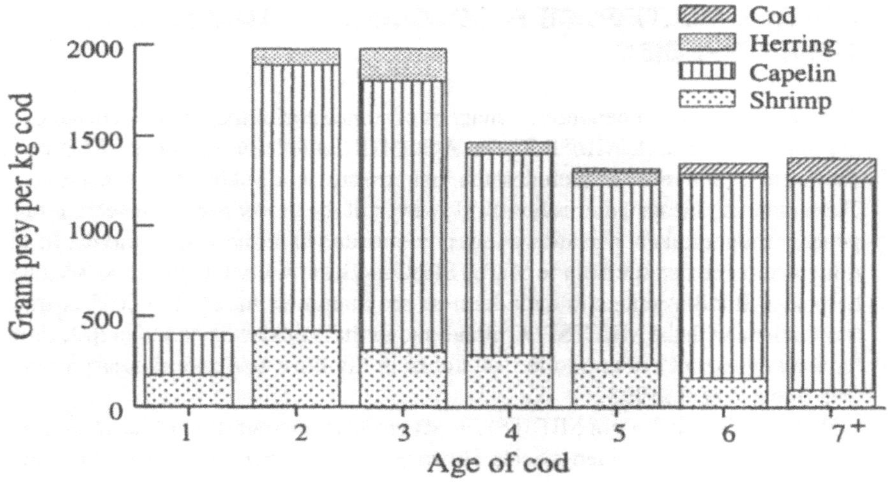

Figure 10. Arcto-Norwegian cod's age dependent average annual consumption of some commercially important prey species. Species included are shrimp (*Pandalus borealis*), capelin (*Mallotus villosus*), herring (*Clupea harengus*) and cod (*Gadus morhua*) above 5, 10, 10 and 20 cm, respectively. In gram prey per kg cod, 1984-92. Sources: Calculations based on data from The Institute of Marine Research, Bergen.

Figures 10 and 11 above indicate that shrimp should be included in the bioeconomic multispecies analysis of the Barents Sea fisheries. Whether this should be done as an integrated part of AGGMULT and MULTSPEC, or it is sufficient to have shrimp as a submodel with cod affecting shrimp, but not vice versa, is left to the biologists to decide.

Minke whale - fish interactions

Flaaten, 1988 is a bioeconomic multispecies analysis of the Barents Sea area fisheries, using a highly aggregated three species model of sea-mammal - fish - fish interactions. One of the conclusions was that the fish preying sea mammals were costly to the fishing industry and therefore should be heavily exploited to increase overall resource rent from the ecosystem. Multispecies models that include interactions between sea mammals and fish are not often available in a suitable form for bioeconomic analysis. Therefore, Flaaten and Stollery, 1996 developed methods for the calculation of the net cost per unit which predators inflicted upon the prey species fisheries. Applying one of the methods to the North East Atlantic stock of minke whales predation of fish, they estimated the annual predation costs per minke whale. The following paragraphs and results are from Flaaten and Stollery, 1996.

In 1989 the stock estimate of the NE Atlantic minke whale was 86,736 whales with a 95% confidence interval of between 60,736 and 117,449 (IWC, 1992).

The energy requirement of a single whale depends on several physical and physiological factors, of which the most important are: weight, basic metabolism, growth,

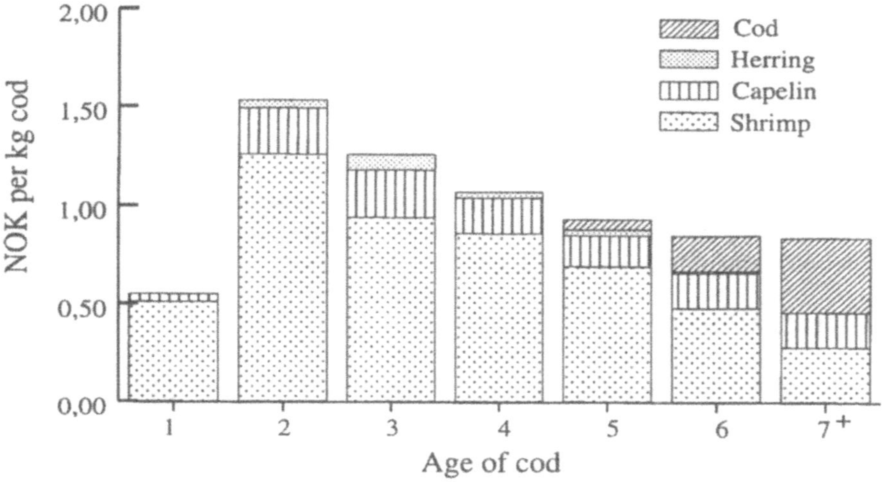

Figure 11. Age dependent average annual opportunity cost of Arcto-Norwegian cod's consumption of some commercially important prey species. Species included are shrimp (Pandalus borealis), capelin (*Mallotus villosus*), herring (*Clupea harengus*) and cod (*Gadus morhua*) above 5, 10, 10 and 20 cm, respectively. In NOK per kg cod, in 1991-92 prices. Consumption data from 1984-92. Sources: Calculations based on biological data from the Institute of Marine Research, Bergen, and economic data from The Directorate of Fisheries, Bergen.

reproduction, heat loss to the environment, swimming activity for foraging, movement between food patches and seasonal long-distance migration between tropic/subtropic waters and the arctic feeding grounds. In addition, the energy acquired in the NE-Atlantic feeding grounds also depends on the amount of energy stored in the blubber, and on the efficiency of the whale's transformation of the gross energy content of the prey into metabolizable energy. Estimates of energy requirement in whales and seals are often based on indirect methods where heart capacity and respiration play an important role. For the NE-Atlantic minke whale Markussen et al. (1992) have estimated the energy requirement using a physiologically-based simulation model. A stock level of 86,700 minke whales would have a total annual energy requirement of $4.166 * 10^{12}$ kcal, or a consumption of 2.314 million tonnes of food with an average energy density of 1800 kcal kg^{-1}. This corresponds to 26.7 tonnes per minke whale per year. Estimates of the energy and food requirement for parts of the year are for obvious reasons lower (see e.g. Nordøy et al., 1995).

Compared to most other baleen whales the minke whale has a varied diet which includes economically important species of fish. Investigation of the stomach content of slaughtered animals is the traditional method to determine the prey species. In addition to herring, capelin and cod, other important commercial species are haddock and saithe.

Flaaten and Stollery (1996) estimated the annual predation costs per minke whale at between ECU 1,440 and ECU 1,910. Their estimates were based on three different diets, two sets of energy content of prey and Norwegian price of fish and cost of effort data. The contribution margin per unit prey equals 30% of the quay side price of fish. It is important to note that the winter diet of the NE Atlantic minke whale is not known, and that the predation cost estimates by Flaaten and Stollery (1996) are based on the assumption that the winter diet equals the known diet from the rest of the year. Even if it is not known where the NE Atlantic minke whale feed during the winter months, a significant part of the predation costs are inflicted upon the fishing industries of Norway, Russia and other European countries.

CONCLUSION

Multispecies modelling is useful, if not necessary, for improved overall management. This is especially so when there are strong predator-prey, or competitive biological interactions among species that can be harvested independently of each other. A biological multispecies model gives the biological restriction on the possible combinations of harvest rates for the species in a particular area. In addition, a bioeconomic multispecies model helps to pick the optimal combination of harvest rates.

In many cases, multispecies biological modelling helps to improve our general understanding of the ecosystem and its fisheries. On the other hand, modelling and data collection by research vessels is not at all costless. Perhaps the economists' initial contribution in some cases should be simpler cost-benefit analysis of marine resources, to get a first idea of whether it is likely or not that more complex and detailed models can improve management. The cases of cod - shrimp and minke whale - fish analysis above are such examples. However, if a biological multispecies model already exists, it is probably a good investment for society to let resource economists extend it into a bioeconomic multispecies model.

Biological research strongly indicate that fish consuming sea mammals may have an effect on the fish stocks of the Barents sea (see e.g. Bogstad et al. 1995, Markussen et al.,1992). Also, bioeconomic research strongly indicate that the predation effects on fish from sea mammals in this area may be of considerable economic importance for the fishing industries (see e.g. Flaaten,1988; Flaaten and Stollery, 1996). Therefore, sea mammals, in particular minke whale and harp seal, should be included in future versions of AGGMULT and ECONMULT.

The biological and economic effects on the shrimp stock and the shrimp fisheries from cod's consumption of prey was briefly discussed in the previous section. The results strongly indicate that there may be a negative effect from the cod stock on the shrimp fishing industry. Therefore, the shrimp stock should be included in the multispecies model AGGMULT, and the shrimp fisheries be included in the multifleet model ECONMULT.

The simulations presented in this paper use all 24 cod fishing vessel groups in all simulations, regardless of their cost effectiveness. Future works should include more thorough economic analysis of the different vessel groups and their performance under different biological and economic conditions. This also applies to the capelin fishing vessels.

For the development of the multifleet model ECONMULT, to be used with the biological multispecies model AGGMULT, it has been necessary with close co-operation between economists and natural scientists. The ECONMULT simulations shown in this paper are not the final product of this co-operative work, but rather one more step on a long way of further research.

REFERENCES

Bogstad, B., K. Hiis Hauge & Ø. Ulltang 1995. Results of simulation studies using a multispecies model for the Barents Sea (MULTSPEC). Int. Whal. Commn., SC/47/NA1, 33 pp.

Eide, A., 1991. ECONSIMP, Econmult-beregninger basert på biologimodellen Multsimp. Working paper. In Norwegian, June 1991. Norges Fiskerihøgskole. Tromsø, 36 pp.

Eide, A. 1992. Econmult 3.0, Manual. In Norwegian. Norges Fiskerihøgskole, Universitetet i Tromsø, 14 pp.

Eide, A. 1993. ECONMULT/ECONSIMP2. Bioøkonomisk flerbestandsmodellering for fiskeriene i Barentshavet. In Norwegian. Norges Fiskerihøskole, Universitetet i Tromsø, 18 pp.

Eide, A. 1994. Multispecies Fisheries Management. A Bioeconomic Multispecies Multifleet Simulation Model of the Barents Sea. Paper presented at the Seminar on Intertemporal Resource Management, Oslo, August 1994, 20 pp.

Eide, A & O. Flaaten 1994. Bioeconomic multispecies modelling of the Barents Sea fisheries, Pp. 183-191 (Vol. I), in: M. Antona, J. Catanzano & J.G. Sutinen (eds.). Proc. Intern. Inst. of Fish. Econ. and Trade's Conf. July 6-9, 1992, Paris. Issy-les-Moulineaux. IFREMER.

European Communities 1987. Assessment of technical interactions in mixed fisheries. Commission of European Communities. Brussels.

Flaaten, O. 1988. The economics of multispecies harvesting - theory and application to the Barents Sea fisheries. Studies in Contemporary Economics. Berlin - Tokyo. Springer-Verlag.

Flaaten, O. & K. Stollery 1996. The economic costs of biological predation - theory and application to the case of the North-East Atlantic minke whale's (Balaenoptera acutorostrata) consumption of fish. Environ. resou. econ. 8: 75-95.

Haug, T., U. Lindstrøm, K.T. Nilssen, I. Røttingen & H.J. Skaug 1995. Diet and food availability for Northeast Atlantic Minke whales, Balaenoptera acutorostrata. Int. Whal. Commn. SC/47/NA2.

ICES 1993. Reports of the ICES advisory committee on fishery management, 1992. ICES coop. res. rep. no.193. Copenhagen.

IWC 1992. Report of the Scientific committee. Rep. Int. Whal. Commn. 42.

Lotka, A.J. 1925. Elements of physical biology. (Reissued as Elements of mathematical biology by Dover, 1956. Baltimore. Williams & Wilkins.)

Markussen, N.H., M. Ryg & C. Lydersen 1992. Food consumption of the NE Atlantic minke whale (Balaenoptera acutorostrata) population estimated with a simulation model. ICES J. Mar. Sci. 49: 317-323.

Mauchline, J. & L.R. Fisher 1969. The biology of euphausiids. Adv. Mar. Biol. 7: 1-454.

Mehl, S. 1989. The Northeast Arctic cod stock's consumption of commercially exploited prey species in 1984-1986. Rapp. P.-v. Reun. Cons. Int. Explor. Mer. 188: 185-205.

Nilssen, K.T. 1995. Seasonal distribution, condition and feeding habits of Barents Sea harp seals (Phoca Groenlandica). Ph.d. thesis, The Norwegian College of Fishery Science, University of Tromsø, Tromsø.

Nordøy, E.S., L.P. Folkow, P.E. Mårtensson & A.S. Blix 1995. Food requirements of Northeast Atlantic minke whales. Pp. 307-317 in: A. Schytte-Blix, L. Walløe & Ø. Ulltang (eds.). Whales, seals, fish and man. Amsterdam, Elsevier.

Tjelmeland, S. 1990. MULTSIMP: A simplified multispecies model for the Barents Sea. Working paper. Institute of Marine Research. Bergen.

Tjelmeland, S. 1992. AGGMULT. A documentation. Internal report, IMR, Bergen.

Tjelmeland S. & B. Bogstad 1989. MULTSPEC: The manual. Institute of Marine Research. Bergen.

Verhulst, P.F. 1838. Notice su la loi que la population suit dans son accrossement. Corr. Math. Phys. 10: 113-121.

Volterra, V. 1928. Variations and fluctuations of the number of individuals in animal species living together. J. du Cons. III: 1-51. (Conseil Permanent International pour L'Exploration de la Mer).

Zenkevitch, L. 1963. Biology of the seas of the USSR. London. George Allen & Unwin Ltd.

APPENDIX

Table A1. Maximum number of vessels available during the year in different fleet groups, fixed costs (on quarters), variable costs of each unit of effort (vessel) and stock output elasticities of fleet groups. (For fotnotes see next page).

Vessel group no.	Type of gear	Vessel length (m.o.a.l.)	Max. no. of vessels[1] Quarter 1	Quarter 2	Quarter 3	Quarter 4	Costs Fixed quarterly[2] ('000 NOK)	Variable[3] (NOK.day-1)	Stock-output elasticity[4]
1	Purse seine,summer	>48.9	0	0	103	103	550	12,347	0.50
2	Purse seine	>48.9	103	0	0	103	550	12,347	0.50
3	Pelagic trawl	43.0-48.9	36	0	0	36	127	3,082	0.50
4	Purse seine	19.0-21.9	36	0	0	36	53	1,099	0.50
5	Purse seine	22.0-24.9	36	0	0	36	83	1,895	0.50
6	Gillnet	<16.0	2,100	1,050	420	630	33	777	0.73
7	Gillnet	16.0-18.9	420	210	84	126	46	1,827	0.73
8	Gillnet	19.0-21.9	213	106	43	64	99	2,590	0.73
9	Gillnet	22.0-24.9	95	48	19	29	127	2,908	0.73
10	Gillnet	25.0-30.9	56	28	11	17	111	4,219	0.73
11	Longline	<16.0	515	258	103	155	39	2,043	0.35
12	Longline	19.0-21.9	56	28	11	17	67	4,791	0.35
13	Autolongline	<16.0	22	11	4	7	60	1,164	0.35
14	Autolongline	16.0-18.9	17	8	3	5	38	3,637	0.35
15	Autolongline	31.0-36.9	230	115	46	69	112	8,372	0.35
16	Autolongline	37.0-42.9	73	36	15	22	193	8,752	0.35
17	Danish seine	<16.0	336	168	67	101	30	820	0.44
18	Danish seine	16.0-18.9	179	90	36	54	46	976	0.44
19	Danish seine	19.0-21.9	134	67	27	40	72	2,072	0.44
20	Danish seine	22.0-24.9	56	28	11	17	77	1,581	0.44
21	Danish seine	25.0-30.9	11	6	2	3	66	1,599	0.44
22	Bottom trawl	22.0-24.9	7	7	7	7	96	2,511	0.50
23	Bottom trawl	25.0-30.9	13	13	13	13	102	3,992	0.50
24	Bottom trawl	31.0-36.9	13	13	13	13	389	8,701	0.50
25	Bottom trawl	37.0-42.9	14	14	14	14	22	16,238	0.50
26	Bottom trawl	43.0-48.9	14	14	14	14	1,030	17,379	0.50
27	Bottom trawl	>48.9	13	13	13	13	3,358	31,390	0.50
28	Hand-line	<16.0	1,736	868	347	521	19	423	0.44
29	Hand-line	16.0-18.9	34	17	7	10	25	515	0.44

1. In simulation 8 the maximum number of vessels in all quarters are equal to that of quarter 1.
2. From Fasmer (1990).
3. Calculated from Fasmer (1990), using the calculation procedure of Eide and Flaaten (1991).
4. The stock-output elasticity of trawl and purse seine fishing is assumed to be 0.50. The other values are from Flaaten (1987).

Table A2. Ex-vessel prices of fish (Nkr./kg, 1989 prices).

Type of fisheries and gears		Vesselgroups no.	Price (Nkr./kg)
Codfisheries	Other gear	1-16, 24-25	7.26
	Bottom trawl	17-22	4.73
Capelin fisheries		26-30	0.60[1)]
			0.43[2)]

1. For consumptions.
2. For reduction.

Table A3. Catchability coefficients (q 10^3) of the different fleet groups of the cod fisheries, separated on 7 cohorts. The calculation method is described in Eide (1993).

Gr.	3	4	5	6	7	8	9+
6	0.0000	0.0000	0.0008	0.0024	0.0029	0.0025	0.0011
7	0.0000	0.0000	0.0017	0.0051	0.0061	0.0052	0.0024
8	0.0000	0.0000	0.0031	0.0090	0.0108	0.0092	0.0042
9	0.0000	0.0000	0.0033	0.0097	0.0117	0.0100	0.0045
10	0.0000	0.0000	0.0032	0.0093	0.0111	0.0095	0.0043
11	0.0026	0.0170	0.0359	0.0440	0.0046	0.0000	0.0000
12	0.0037	0.0242	0.0510	0.0625	0.0066	0.0000	0.0000
13	0.0024	0.0159	0.0336	0.0412	0.0043	0.0000	0.0000
14	0.0051	0.0335	0.0707	0.0866	0.0091	0.0000	0.0000
15	0.0109	0.0717	0.1515	0.1857	0.0195	0.0000	0.0000
16	0.0127	0.0838	0.1770	0.2170	0.0228	0.0000	0.0000
17	0.0006	0.0028	0.0061	0.0149	0.0097	0.0050	0.0028
18	0.0012	0.0052	0.0113	0.0277	0.0182	0.0094	0.0051
19	0.0015	0.0065	0.0143	0.0351	0.0230	0.0119	0.0065
20	0.0014	0.0059	0.0129	0.0317	0.0208	0.0107	0.0059
21	0.0013	0.0056	0.0122	0.0298	0.0196	0.0101	0.0055
22	0.0023	0.0089	0.0213	0.0438	0.0304	0.0169	0.0098
23	0.0031	0.0117	0.0281	0.0576	0.0400	0.0222	0.0129
24	0.0067	0.0258	0.0617	0.1268	0.0880	0.0489	0.0285
25	0.0088	0.0336	0.0804	0.1652	0.1146	0.0637	0.0371
26	0.0094	0.0359	0.0859	0.1765	0.1224	0.0680	0.0396
27	0.0164	0.0627	0.1502	0.3087	0.2142	0.1190	0.0693
28	0.0007	0.0044	0.0089	0.0100	0.0011	0.0000	0.0000
29	0.0004	0.0027	0.0054	0.0060	0.0007	0.0000	0.0000

Scenario Barents Sea
A Tool for Evaluating Fisheries Management Regimes

By Gro Hagen[1], Einar Hatlebakk[2] and Tore Schweder[3]
[1] Norwegian Computing Center, Box 114 Blindern, N-0314 Oslo, Norway.
[2] Norwegian Government Computing Centre, Box 6664 Rodeløkka, N-0502 Oslo, Norway.
[3] Department of Economics, University of Oslo, Box 1095 Blindern, N-0317 Oslo, Norway.

ABSTRACT

The issues of scenario modelling in general are discussed. A scenario model for the Barents Sea fisheries is presented, with VPA based management for cod and herring, and capelin management based on the CapTool procedure.

Probing scenarios are used to investigate specific questions, and simulations describing selected issues connected to the management procedures are presented.

Uncertainty scenarios are used to evaluate one management procedure, or to compare several, and issues of robustness and efficiency are central to the analysis of simulation results. An uncertainty experiment, with scenarios chosen purely for demonstration, is simulated, and the main results are presented. With mean yearly catches as the only response, the best management strategy is found to be a combination of F_{high} for cod, a spawning target of 500 000 tonnes for capelin, and a constant nominal fishing mortality rate of F=0.20 for herring. When considering biomass variables as additional performance measures, it is found that the 500 000 tonnes spawning target still is the best for capelin, but the cod procedure should rather be F_{med} than F_{high} and the herring procedure should be the yield-per-recruit based strategy $F_{0.3.}$

INTRODUCTION

The revised management procedure of the International Whaling Commission (Anon., 1993) was developed through a long-term process of what we will call scenario modelling. Scenario modelling is an attractive methodology for evaluation and comparison of management regimes, also for other fisheries than whaling. The present paper is an attempt at explaining and developing the concepts and methods of scenario modelling.

The paper presents the achievements of the first two years of a project aiming at investigating management regimes in the Barents Sea. There are three interacting species in the system: cod, capelin and herring. The fisheries are managed by quotas, but the fishing mortality, and the catches, are to some degree influenced by economic considerations. To span the region of uncertainty with respect to population dynamics, quality of survey data and VPA methodology, and economics, we have defined 43 scenarios for 10 binary factors and 30 variants of combined management regimes for the 3 species. Each scenario is run over 90 years, and in 3 replicates. The simulation output is analysed by linear methods, and the management strategies are characterized with respect to robustness and efficiency.

The present study has benefited from the expertise of a group of researchers in marine ecosystem modelling, fisheries economics and management. The development of a simulation model and of scenarios that span the region of plausibility with respect to biology and economics is, however, a demanding task, and the present project has not developed to maturity yet. The results presented should be understood as illustrative of the concepts and methods of scenario modelling in fisheries, and not as definite results directly relevant for Norwegian fisheries management.

SCENARIO MODELLING IN GENERAL

Model experiments and other parallels to scenario modelling

The science of fisheries management is a pragmatic science with many parallels to engineering. The core in fisheries management is a management regime (or strategy) for a particular fishery (group of fisheries). The aim is to regulate the fisheries so as to obtain a good continuing yield. The management strategy is composed of methods for estimating abundance and other parameters in the management model for the ecological system at hand, and a control law for calculating quotas or other regulating quantities when the management model has been estimated on the incoming data. The management strategy is parallel to a design for physical constructions (say ship hulls) in engineering. Engineering designs are supposed to perform satisfactory according to more or less formalized performance criteria. As in fisheries, knowledge of the system in which the engineering construct is to operate will be imperfect. It will be required that the performance is reasonably robust with respect to uncertainty attached to system behaviour. It is also required that the design/management strategy should be able to handle the better understood statistical variability (natural variation) of the system.

Model experiments are an important tool in engineering. In ship building, small scale models of hull designs are tested in a basin where waves, wind and current are manipulated according to a research plan. The scenarios for the real ocean system simulated in the model basin will typically span the range of plausible states of nature. Repeated simulations will be needed to obtain statistical information on performance under natural variability inherent in each scenario. A rigid observational regime is imposed to register the performance of the model ship under the various scenarios in replicated trials. To draw the parallel, the simulation model corresponds to the test basin and the management strategy corresponds to the ship hull to be tested.

Scenario modelling in fisheries science

Physical experiments are not feasible for testing out management regimes for fisheries regulation. For large systems like marine ecosystems, computerbased simulation modelling is the only practical technology for model experiments of management regimes. This will in fact also be the case in various fields of engineering and other pragmatic sciences, and it is indeed the case for policy experimentation in economics. We will use the terms *scenario modelling* for the process of establishing a computerbased test bench for management regimes for fisheries, and *scenario experimentation* for carrying out simulation experiments to evaluate a given management strategy. A scenario model has three main ingredients: a *simulation model*, a *management procedure* and a set of *performance measures*.

Our *simulation model* is a parametric model capable of representing the main dynamics of the ecological system in the Barents Sea, and its interactions with human activity like fishing and resource surveying. The simulation model should provide a conceptually simple computerbased framework, which allows the competent researchers to specify the various plausible states of nature, including human activity, which they wish to consider. The set of such scenarios should ideally span the uncertainty with respect to knowledge of the system. Uncertainty in the form of natural variation is incorporated in each scenario as stochastic variation.

The *management procedure* is a feedback mechanism representing a management regime for the system. The regime in current use is a single species VPA analysis of cod and herring, with capelin quotas to target the spawning stock at a given level and allowing the cod to have its estimated demand for capelin. As other management regimes, this is a mixture of abundance estimation of the relevant stocks, based on data from the fisheries and the surveys, and of a control law for calculating quotas on basis of the abundance estimates. To represent this regime in a *management procedure*, the methods behind the various informal judgements involved must be clearly specified. A *management procedure* is thus a computerized formalization of a management regime which interacts with the simulation model to constitute a fully specified dynamic model for the system. The scenario model will be made capable of handling various management regimes to be evaluated.

The evaluation of a management regime will be based on appropriate *performance statistics*. These statistics represent features of the simulated development of the system for a given scenario, and for the management procedure under study. Summary statistics based on the joint distribution of performance statistics like mean yearly catch of cod, value of total catch, spawning stock of cod at end of period etc., over replicated runs of the same scenario, will be used to characterize the management regime.

Scenario modelling and *experimentation* may be used for various purposes. One purpose is simply to investigate the performance of a given management regime, say that presently in use for the Barents Sea fisheries. Because of the uncertainty attached to our understanding of the ecological system and its interaction with fishing and research, several scenarios must be considered to reasonably span the plausible range. To delineate this plausible range of scenarios is a valuable exercise in itself. Features common to the various scenarios will represent knowledge of the system, and variability in features across scenarios will represent the uncertainty attached to the knowledge. Ideally, it should be the knowledge and uncertainty as agreed by the scientific community of researchers in relevant fields of oceanography, marine biology, fisheries science etc. which should be represented by the set of scenarios. When disagreement and uncertainty are great, many and diverse scenarios will be needed. The process of establishing a set of scenarios will usually have to be a co-operative process of a group of scientists from the various relevant fields and laboratories.

In addition to the *simulation model* and its agreed set of scenarios, the management regime will need to be formalized into a *management procedure*. To make explicit how fisheries actually are managed makes this field of policy more open for insight and debate than is the case when judgements and estimation are done more informally. This increased openness will hopefully enhance improvement in the management strategy.

The third component of the *scenario model* for testing out a current management regime, the set of *performance statistics* and their internal weighting, will bring to the open a debate over the aims and goals of fisheries management. The various fractions of the fisheries industry, as well as fisheries economists, planning bureaucrats and politicians, will all have ideas of performance criteria by which to judge a management regime. These ideas are not necessarily in agreement. To get a reasonable set of *performance measures*, co-operation between various interest groups will be required. The task of hammering out a minimal set of explicit *performance statistics*, on which to base the evaluation of a management regime, will hopefully stimulate debate and co-operation between interested parties, and even possibly result in an agreed overall criterion based on balancing the often mutually competing sub-criteria.

In summary, the process of scenario modelling and experimentation to investigate the current management regime, and compare it to other possible regimes, will hopefully represent a challenge, and provide a research environment which will crystallize the scientific knowledge and uncertainty of the ecological and fisheries system at hand, open up the actual management regime for scrutiny and improvement, and enhance the understanding of the aims and goals for management of the given fishery. Further use of *scenario modelling* will then be to identify important research problems, improving the management regime, and in helping interest groups, industry and politicians to better understand possibilities and limitations in fisheries management.

Scenario modelling is not commonplace in fisheries science. A pioneering case of scenario modelling was, however, carried out by the Scientific Committee of the International Whaling Commission (IWC). The goal was to develop a management procedure for whaling, which should be robust against the great uncertainty attached to our understanding of the population dynamics of whale stocks and to the great imperfections in abundance estimates and other data on the status of the whale stocks. The work was organized with five different groups of scientists engaged in designing candidate management procedures. Together with a few additional scientists, they met for workshops once every half year over a 5 year period. At the workshops, results were compared and discussed. In addition to discussing the results of the preceding half years development and simulation, the workshop discussed and decided upon further scenarios to be implemented and simulated, and also further development of the performance measures. This forward looking provided a plan for the work of the next half year, and it provided explicit challenges for the management strategy.

An interesting feature of the IWC work is that the procedures to little extent were developed by the use of theory of optimal control. The candidate strategies were rather developed by a process of creative construction, based on simulation results and exchange of ideas between the research groups. This was partly due to the fact that emphasis was on robustness versus imperfection in knowledge and data, and not on optimization.

The candidate procedures did not converge in structure or method. To a considerable degree, however, they did converge in performance. By tuning the candidate procedures to common targets in a reference scenario, it was possible on the basis of simple descriptive statistics computed from the performance measures, to conclude that one of the procedures was to be preferred and that its robustness properties were acceptable for real implementation. The revised management procedure is described by IWC in Anon. (1993) and by Cooke (1995).

Different types of fisheries models

It is helpful to distinguish between three basic types of models: assessment models, management models and scenario models.

The purpose of an assessment model is to organize the information concerning one stock, or an ecosystem consisting of several stocks, so as to obtain the best possible assessment of the status of the stock, or the system. Assessment models are often complex, and the issues of parameter estimation are central when building and using an assessment model. The various VPA models and methods are cases of assessment models. Hilborn (1992) reviews the state of the art in stock assessment for the purpose of management. He makes the point that these various methods generally rely on one set of data each, ignoring other relevant sets of data. In fisheries, various models of the multispecies type have been developed for the purpose of assessment. The MULTSPEC model by Bogstad & Tjelmeland (1993) has been developed for assessing capelin and cod in the Barents Sea. In the Scenario Barents Sea project, we draw on the experience and competence of the MULTSPEC group, and in due time, management strategies based on MULTSPEC might be tested out in the Scenario Barents Sea model.

A management model is a feedback model aimed at explicating the management strategy. The management model will often be a rather simple assessment model coupled with a control law.

As explained above, scenario models are, so to speak, model laboratories for testing out assessment models and management models (or rather, assessment methods and management schemes).

Probing and uncertainty scenarios

In the Scenario Barents Sea project, scenarios are organized in two groups. The *probing scenarios* are chosen to investigate specific questions, often one at the time. They are selected to probe current management issues, but also to improve our understanding of the model. *Uncertainty scenarios*, on the other hand, are systems of scenarios designed to span a plausible region in state space. A reference group of experts in fishery biology, economy and management is instrumental in setting plausible limits to parameters and structures in the simulation model. Hypothesis generated by the probing scenarios might be tested in uncertainty scenarios. Issues of robustness and efficiency are central to the analysis of simulation results for uncertainty scenarios.

A SCENARIO MODEL FOR THE BARENTS SEA

As already mentioned, the scenario model has three main ingredients, the *simulation model*, the *management procedure* and the *performance measures*. This section presents these three ingredients briefly, as they are currently implemented.

The simulation model

Our model is supposed to be a rather simple simulation model of the ecosystem in the Barents Sea, containing the main biological and environmental factors. The intention is not to build a "true" model, but to establish a framework for operational specification of various plausible hypothesis concerning the fisheries system. As such, it must be capable of producing variability at the same level as observed in nature. The main factors in the simulation model will depend on a few essential parameters. Ideally, it should be possible to imitate plausible states of nature (ecosystem and fishermen's behaviour) by varying these parameters.

In order to make the simulation model practical and understandable, the concepts must be simple in structure and nature. The more it is based on biological and economic theory, the better. Only the main factors in the system will be explicitly modeled. At present, these are:

- Three species of fish: cod, capelin and herring. All three species are modeled as age- and length distributed. Recruitment, maturation, growth, predation and natural mortality for all three species are modeled as parameter dependent. Both the parameters and the functions describing these mechanisms can be changed.

- Human activities in the Barents Sea: fishery and research surveys. In the simulation model, fishery is carried out on the simulated stocks according to quotas given by the management procedure. There is room for overfishing and discards of small fish. Data from the research surveys are modeled when they are needed as input to the management procedure. At present we have implemented only acoustic stock estimates for herring and capelin, and tuning indices for the cod stock. As for catch based input data to the management procedure, plausible levels of bias and variability are imposed on the survey data.
- Some environmental factors, like sea temperature and primary production. At present, these environmental factors are subsumed in a few proxy variables which generates a random environment for the interacting fish stocks.

The management procedure itself is not a part of the simulation model, and may be replaced by other management procedures. The simulation model communicates with the management procedure by giving necessary data input to it, and by letting the data fisheries be regulated by the quotas set by the management procedure.

The main elements in the model system are the fish stocks and human activity related to these. The purpose of the scenario model is to follow the stock abundances and the catches over a period of time under specified environmental conditions and a given management regime. During a year the stocks are exposed to reducing factors (natural mortality, predation, fishery) and to increasing factors (growth, recruitment), and the simulation model is supposed to be capable of imitating all these factors in various plausible ways. Natural variations in the system are mainly modeled by probability distributions.

Before turning to more details, let us shortly describe the dynamics in the simulation system. The system starts at 1 January after being initialized. Then recruitment is computed and added to the stocks, and the age distributions are updated. The appropriate components of the cod and herring stocks are maturing. The next step in the process is to compute quotas for each stock, by calling the management procedure. For each month following, mortality and growth are executed (in that order). When arriving at 1 October, capelin is maturing and so divided in two separate stocks. The age structure is discrete, and age is incremented at the turn of the year for all three species.

Growth

Increase in individual length, L, is modeled with a difference equation :

$$\begin{aligned} dL(t) &= L(t+\Delta_t) - L(t) \\ &= (L_\infty - L(t)) \cdot G(t) \cdot M(t). \end{aligned} \tag{1}$$

$M(t)$ is a discrete function, taking values in the interval [0,1], and the function is used to model monthly variations in growth over the interval ["no growth", growth potential"] (a seasonal profile). In the simulations presented, the following monthly profile values are used ($t = 1, 2, \dots, 12$):

Cod:	$M(t) \equiv$	1, all months (no seasonal variation)
Herring:	$M(t) =$	$(0,0,0,0,1,1,1,1,0.5,0.25,0,0)$
Capelin:	$M(t) =$	$(0,0,0,1,1,1,1,1,1,0,0,0)$

The growth-function, G, for cod depends on the predation, and is explained in detail in (Hagen *et al.*, 1995) and more briefly in the appendix. For capelin and herring the model is

$$G(t) = 1 - \exp(-K(t)\Delta_t), \tag{2}$$

and the parameters are explained in Table 1. Increase in length is computed monthly. Mean weight in each length group is preliminary given as parameters to the simulation model, and these weights are not affected by growth. This entails that the model at this stage not is capable of handling emaciation due to lack of food. However, absence of suitable food results in a slower length growth. Emaciation due to spawning is taken care of, and for the mature part of the stocks mean weight is reduced after spawning. During one year the normal mean weights are regained in all length groups, however.

Recruitment

In the simulation model, juveniles are recruited into the stocks between age 1 and 2, capelin as one year olds on 1 October, cod and herring as two year olds on 1 January.

To describe the relation between the mature stock and the number of recruits some months (or years) later, we use the Beverton-Holt function:

$$R_t = MAXR \cdot \frac{B_t}{B_t + H} \tag{3}$$

Table 1. The parameters involved in the functions (1) and (2). All parameters, exept the sea temperature and the time variables, are specific for each species of fish.

Parameter		Description
t	=	time variable.
Δ_t	=	time interval of fixed length. (one month in the simulation model).
L_∞	=	maximum length.
$K(t)$	=	$(a+be^{-B(t)})\, g(\Delta T, f)$.
		a,b = parameters.
		$B(t)$ = stock abundance at time t, (indicating density dependent growth).
		g = function to regulate growth according to sea temperature. Linear in ΔT.
		ΔT = temperature deviation from a defined "normal" value.
		f = index of food availability.

This function gives the expected number of recruits, R_t, as a function of the mature biomass, B_t. The parameters describing maximum level (*MAXR*) and half value (*H*) are of course specific for each species of fish. The natural variation is modeled as stochastic white noise. However, from the historical data it is evident that there are years with extremely high recruitment. This is modeled by having two levels of recruitment, one "normal" level and one "high" level. The "high" level occurs when there are favourable survival conditions, e.g. favourable sea temperature. The two levels are both modeled as Beverton-Holt functions, but with different parameters.

Instead of explicitly linking sea temperature to recruitment level, we have usually chosen to model stochastic variation between the "normal" and the "high" recruitment level. The expected waiting time between two years of "high" recruitment is approximately p^{-1} years. Since there hardly have been periods with two or more years of high recruitment in sequence in the Barents Sea, a period of normal years is imposed after a year of "high" recruitment. After a year of "high" recruitment a period of $(2p)^{-1}$ years passes with normal recruitment. Thereafter the recruitment level is drawn at random, but with a probability of $2p$ of drawing the "high" level ($p<0.5$). In this way, the expected waiting time between two years of extreme recruitment still is p^{-1} years, but the waiting time period will never become less than $(2p)^{-1}$ years. The "success probability" p is a system parameter, and can be changed. The years of "high" recruitment occur simultaneously for all three species.

From the historical data one can argue that the waiting time period between two years of extreme recruitment is approximately 10 years, which corresponds to a p-value in the simulation model of 0.1. We have chosen to use this value in the simulations.

If simulation is performed with temperature effect, the sea temperature will also have influence on the recruitment. This is done by regulating the maximum number of recruits according to the temperature. High temperature is favourable and more juveniles will survive until the age of recruitment. The regulating function is linear in ΔT (temperature deviation from the defined "normal" temperature), and is identical for all three species. If "high" recruitment coincides with high abundance of (young) herring in the Barents Sea, increased predation from herring on capelin might counterweight the recruitment effect, and the result might be reduced capelin recruitment. This factor is modeled similarly to the temperature influence, by a linear function lowering the maximum number of capelin recruits. Both this influence, and the temperature influence can be "turned off". In the simulations presented in this article, there is no temperature influence included.

Maturation

Maturation is modeled as a pulse. Cod and herring matures at the turn of the year, while capelin matures on 1 October. These dates are parameters to the simulation model, and might be changed. The maturity ogives (fractions of mature) in each length group are logistic functions:

$$m(l) = \frac{1}{1+e^{4P_1(P_2-l)}}, \tag{4}$$

where l denotes mean length in the length group (midpoint in length interval). The two parameters P_1 and P_2 are specific to each species of fish, P_2 being the "median" length at maturation. Maturation is not sex-dependent.

Mortality

There are three types of mortality:

Predation mortality due to species included in the model, mainly cod.
Fishing mortality.
"Natural" mortality, including predation mortality from species not explicitly represented in the simulation model.

Predation mortality

Predation is a major component of the dynamics of the simulation model, and a quite detailed description of the predation model is given in the appendix. In the simulation model cod is main predator and capelin is prey. When present in the Barents Sea, herring is prey for cod, and also possibly predator on small capelin. In addition, cod is some times cannibal, and so the smallest cod is both predator (on capelin and younger herring) and prey (for older/larger cod). The model does not allow stocks to suffer extinction due to predation. When the predator stock is large compared to the prey, reduced individual growth will result.

Fishing mortality

Each month the simulation model computes the total biomass caught. This is done by distributing the year's total catch by a seasonal profile over the year. The resulting biomass is then distributed over the length groups according to the selection pattern in the actual fishery. If C_i is the biomass caught from length group i, and B_i is the total biomass at the beginning of the month, then the fishing mortality this month is given by

$$F_i = \frac{C_i}{B_i}$$

This procedure is followed for each species and length group.

Natural mortality

Natural mortality (excess mortality not covered by the two other types of mortality) is simply modeled as a monthly mortality rate. For capelin, spawning mortality is explicitly modeled, it is possible to parameterize the model so that a given fraction of mature capelin dies after spawning.

Total mortality

If P is the monthly rate of predation mortality, F the rate of fishing mortality, S the rate of spawning mortality and M the rate of mortality due to other reasons, the monthly survival of the stock is:

$$N_{i,t+1} = N_{i,t} \cdot (1-P) \cdot (1-F) \cdot (1-S) \cdot (1-M) \tag{5}$$

where $N_{i,t}$ is the number of individuals in length group i at the beginning of month t.

Fishery and research surveys

Fishery

The cod fishery is divided into several fleets, to be able to model the varying selection patterns from the different gears, and the model can be summarized as follows:

1. Simulated catch-by-age data and survey data are given to the VPA-program, and the model stock is estimated (number-by-age) for the current year.
2. The number-by-age estimate is given to the quota-program, and a recommended total quota, according to the management strategy, is computed.
3. After a check on reasonableness, the resulting quota is sent to the fleet-module. The total quota is divided between the fleets, and a resulting theoretical catch is distributed over length-groups according to the fleet specific selection patterns. Non-reported catches (as a result of general overfishing or discards) are also computed and length-distributed similarly. At last the resulting total selection patterns in reported and non-reported fishery are computed, and these are returned to the simulation program together with the total reported and non-reported quantities.
4. The reported and non-reported fishery in the simulation model are conducted according to these "weighted" selection patterns.

The quotas are computed on 1 January for all three species, and the fishery is distributed over the year according to a seasonal profile. In the simulations presented here, capelin is only subjected to fishery during spring (winter fisheries), while cod and herring are subjected to fishery through most of the year. The seasonal profiles for fisheries are parameters to the simulation system.

From the simulation output it is not possible to distinguish between the general overfishing and the discards in cod fishery. This is of course a drawback, because discards are lost resources, while the general overfishing most often is landed. However, the problem is only present among the smallest length groups. In the larger length groups non-reported catches are always a result from general overfishing and thus landed. For this reason, the catch based performance measures are based on catches in the larger length groups only.

Tuning indices (cod)

Cod assessment depends on data from research surveys, tuning indices, together with age-distributed catch- and weight- data. Each year, 5 research surveys are simulated. The cod surveys are modeled to reflect the model stock at different times of the year: the two first on 15 February, the two next on 1 October and the last one on 1 November. For each survey a total index I is computed from the total stock:

$$I = f(B) + random\ \ noise\ ,$$

where the function f at present is either proportional to B or to $\sqrt{B}$. The random noise is modeled as a normally distributed variable with zero expectation. All parameters (included noise variance) are specific to each survey, and the parameter values are estimated by linear regression on historical survey data and VPA-estimated stock data from ICES' reports (Anon., 1993). At present, the random noise has constant variance. The total index is then distributed over age classes according to the age distribution of the model stock. From real data it is evident that the age distribution of survey indices is not correct, at least not compared to the total stock. The correct age distribution in the simulations is therefore a simplification.

Performance measures

To evaluate a management procedure, or to compare several, statistical performance measures are needed. In the simulations conducted so far, we have used rather simple performance measures. Since cod is the most important species commercially, the response variables have been mainly "cod responses", but in the present simulation study also some simple herring and capelin responses are used. The main response variables in this experiment are

1. Mean yearly catch of cod.
2. Mean yearly catch of capelin.
3. Mean yearly catch of herring.

The means are computed over the total response period. In addition to these variables, also mature biomass for cod and herring are taken into account. To judge a management strategy, we look at the responses on all performance measures included in the analysis. However, it is not obvious that the same strategy will give the best "score" on every response. It will then be necessary to weight the scores from the various responses into a "total score", before a final conclusion can be taken. How to construct such a weighting is not obvious, and is rather a task for biologists and fishery bureaucrats than for statisticians.

Management procedures

Management regimes usually include more or less informal expert judgement in conjunction with statistical analysis of catch and survey data. It is not practicable to include this human element in the modeled management procedures for at least two reasons:

- The procedure will be called many times during a simulation run. A quota is to be set (at least) once a year in the simulation period, and there typically will be

many years (>40), with replications, and also many scenarios to be simulated. It is important that all decisions made during the runs are consistent. By "consistent" we mean that when the conditions are equal, the same decisions shall be made irrespective of scenario or replicate number. Also the time spent on making decisions ought to be minimized.
- It is desirable that the scenario model and its management procedure is explicitly and precisely defined. This opens up for scrutiny, evaluation and improvement, and it might improve the expert judgement that always will have its place in the management of real fisheries.

For each management regime to be evaluated in the scenario model, we therefore need to specify precise instructions for every decision that will be taken. We then impose the required consistency. Even though some knowledge and experience are lost when formalizing the management regime into a management procedure, we hope to minimize the loss by relying on advice from the experts. We will now shortly describe the implemented procedures, and then take a closer look at the management strategies used for cod and herring.

The implemented procedures

Cod

The cod assessment is based on VPA-analysis with Laurec-Shepherd tuning (Laurec & Shepherd, 1983, Pope & Shepherd, 1985). The management strategies implemented are either based on a constant nominal fishing mortality rate, or on biological reference points such as F_{med} or F_{max} (Jakobsen, 1992).

Fishery is modeled according to a length selection, while the quota is calculated according to an age dependent selection. The age selection might be determined from the catches using separable VPA (Pope & Shepherd, 1982), or could be a constant selection pattern computed from the fleets' individual length dependent selection patterns. The structure of the cod procedure is as described in the fishery paragraph in the Fishery and research surveys section.

Capelin

The capelin procedure is based on the CapTool-strategy (Bogstad *et al.*, 1992). In the simulations, fishing is only effected on the mature stock (winter fisheries), and the regulation condition is to maintain a spawning stock of a given size. Predation from cod is taken into account, when the quota is set. Input data to this management procedure are an acoustic estimate of the stock abundance and the assumed cod consumption. This assumed consumption is computed using the VPA estimates for the cod stock.

Herring

The herring procedure is also based on VPA-analysis, but does not involve the tuning procedure. Inputs to the procedure are acoustic stock estimates and reported catches, and quotas are set using biological reference points or a constant nominal fishing mortality. Only reference points based on the yield-per-recruit function are implemented for herring.

Management strategies for cod and herring

The management strategies for cod and herring are represented by nominal fishing mortality rates, F-values, computed by VPA on the basis of the parameters natural mortality, weight, age at maturation and recruitment. The level of fishing mortality is also called a biological reference point.

We consider two types of VPA based management. The first group is based on the ratio between the number of recruits and the corresponding spawning stock biomass, and these strategies are referred to as F_{high}, F_{med} and F_{low}. The second group is based on the yield per recruit function, and the corresponding strategies are referred to as F_{max} and F_p. A more detailed description of all these strategies are found in Jakobsen (1992). Table 2 summarizes these management strategies in a few words.

The strategies F_{high}, F_{med} and F_{low}

F_{high}, F_{med} and F_{low} are estimated from the ratio (R) between the number of recruits and the corresponding spawning stock biomass (SSB). The ratio values are chosen from the ordered, VPA-computed values for R. The median is used for F_{med}, for F_{low} the value of R is chosen where exactly 10% of the observed values are lower, while for F_{high} 90% of the values are lower. Quotas are then computed, based on the ratio number, the natural mortality, individual age, weights and the selection pattern, to match the reference mortality rate.

In our scenario model the F-values are computed based on 25 ratio values, due to limitations in the general output from the VPA-procedure. However, for illustrative computations a longer period might be used, and for cod yearly data are available back to 1946.

We will show an example of the computations of the F-value for the three strategies. The VPA data for cod in 1993 (Anon., 1993) are used in the computations. Figure 1 plots the ratio R as a function of SSB, based on VPA estimates for cod from the period 1946-90. This figure shows that R and SSB are negatively correlated.

Table 2. Management strategies used in simulations.

Strategy	Description
F_{high}	High ratio between recruits and spawning biomass
F_{med}	Mean ratio between recruits and spawning biomass
F_{low}	Low ratio between recruits and spawning biomass
F_{max}	Maximum yield per recruit
$F_{0.1}$	Yield per recruit parameter equal to 0.1
$F_{0.3}$	Yield per recruit parameter equal to 0.3

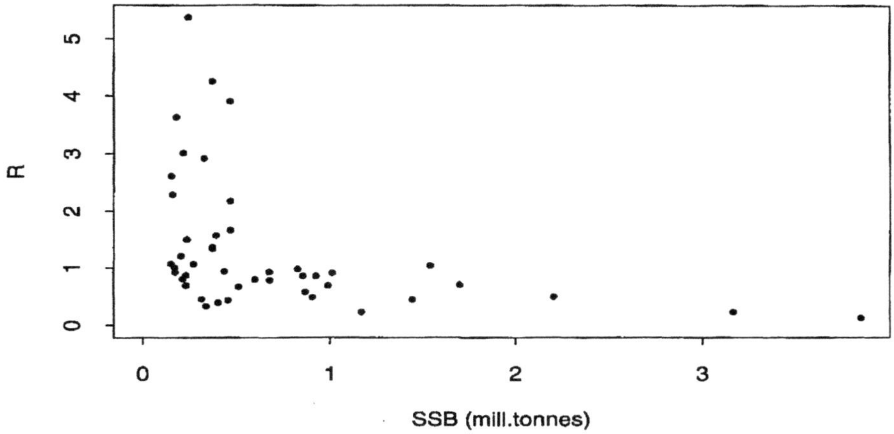

Figure 1. Ratio number (R) of recruits and SSB for cod 1946-1990.

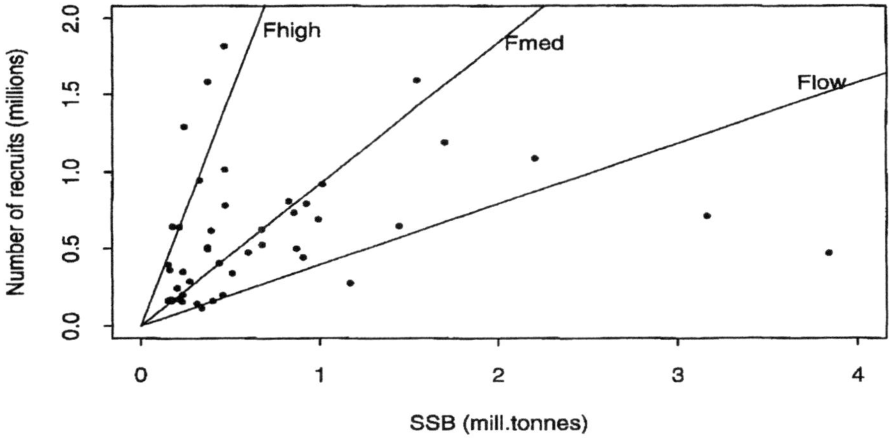

Figure 2. Recruits and SSB 1946-1990, cod.

In Figure 2 the number of recruits, $R \times SSB$, for the period 1946-1990 are plotted against SSB. The lines which determines the ratio number (R) for F_{high}, F_{med} and F_{low} are drawn. The quotas are computed from the corresponding F-values.

In the simulation model, the 1993 quota is determined based on the SSB-values from the period 1964-1988. Table 3 displays the results from the quota computations. The mean F-values and quotas based on the whole period 1946-1993 would have been slightly lower.

The ratio values in Table 3 are higher than the corresponding mean values in the simulations, but the F-values also depend on mean weight in age classes, the size of the mature stock and the selection pattern. Varying these parameters may give other results. In the reference simulations, a ratio of 0.51 gives a mean realized F-value of 0.24 with low growth, while high growth gives a mean F-value of 0.38.

The quotas and F-values computed by F_{high}, F_{med} and F_{low} vary considerably over time. The high values based on the 1993 VPA data illustrate this. F_{med} is designed to keep the spawning stock at the current level, but the quota computed for 1993 is so high that the spawning stock would be reduced.

The computed F-value depends on the natural mortality, and in the reference computations the natural mortality rate are set to M=0.2 in accordance with the nominal natural mortality in the VPA procedure.

The strategies F_{max} and F_p

The management strategy F_{max} maximises the predicted yield per recruit. F_p matches a point where the slope of the yield-recruit curve is p times the slope at the origin ($0 < p < 1$). In Table 4 theoretical 1993 quotas for cod are shown. The strategies F_{max} and F_p are quite stable, but the quotas are lower than corresponding quotas for F_{high}, F_{med} and F_{low}. The quotas and the F-values in Table 4 are nearly the same as the values we get in the reference simulations.

Table 4 also shows theoretical 1993 quotas for herring. The F_{max} quota is very high. It is probably too high to be an interesting strategy for herring. In the simulations the strategy F_p is used with parameter values from 0.1 to 0.3. The natural mortality rate used in the herring computations for Table 4, is age dependent. The M-values are in the interval from 0.13 to 0.23, and are increasing with age.

Table 3. Ratio, F-value and quota in mill. tonnes for cod. VPA data from the years 1964-1988 are used for computing the ratios, and the quota is computed according to the VPA-estimated cod stock in 1993.

Strategy	Ratio	F-value	Quota
F_{high}, 90% point for R	3.007	1.000	0.975
F_{med}, median for R	1.059	0.625	0.705
F_{low}, 10% point for R	0.432	0.386	0.480

Table 4. F-value and quota in mill.tonnes for cod and herring, based on the VPA-estimate in 1993.

	Cod		Herring	
Strategy	F-value	Quota	F-value	Quota
F_{max}	0.267	0.349	2.250	1.777
$F_{0.05}$	0.208	0.280	0.500	0.979
$F_{0.1}$	0.174	0.237	0.375	0.809
$F_{0.2}$	—	—	0.188	0.473
$F_{0.3}$	0.100	0.140	0.125	0.333

The yield per recruit based strategies F_{max} and F_p seem to be more stable than the first group of strategies. These strategies do not intend to preserve the spawning stock. For cod, F_{max} is an alternative to F_{med} etc., and for herring F_p appear to be favourable , $p \geq 0.1$.

Natural mortality dependence

The reference point strategies depend on the nominal natural mortality in the VPA routine. Table 5 shows how the F-values change when this nominal natural mortality rate is varied from $M=0.1$ to $M=0.4$.

For the recruitment/spawning stock based strategies the F-value decreases when the natural mortality increases. F_{med} is designed to keep the spawning stock at a fixed level, and when the natural mortality increases, the fishing mortality is reduced. The total mortality is kept approximately at the same level.

For the yield per recruit based strategies the F-value increases when the natural mortality rate increases in the domain 0.1-0.4. When the natural mortality is high, the maximization of the yield per recruit will produce high catches of young fish with low weight, and so the fishing mortality will be high. When the natural mortality is low, it is favourable to wait until the fish has grown, and consequently a low fishing mortality is set. All strategies depend on weights and fishing pattern (selection pattern), in addition to the natural mortality.

PROBING SCENARIOS

The probing scenarios are not selected according to an experimental plan, but rather on the basis of the current debate over management options. They are also to some degree found by interactive search.

A number of scenarios have been run mainly to study the management procedures: how they react when parameters are varied or some of the assumptions are not correct. In the simulations presented in this section, it is natural mortality and fishing mortality that are varied. The parameters for recruitment and growth are kept fixed. In

Table 5. F-values for cod 1993 with varying natural mortality rate.

	Natural mortality, M-value			
Strategy	0.1	0.2	0.3	0.4
F_{high}	1.156	1.000	0.844	0.680
F_{med}	0.770	0.625	0.481	0.338
F_{low}	0.523	0.386	0.250	0.114
F_{max}	0.200	0.267	0.395	0.656
$F_{0.1}$	0.140	0.174	0.223	0.297

the probing scenarios the management procedures for each of the species are studied separately, with main emphasis on the cod. The results from the simulations are highly dependent on several parameters. The mean stock and quota values that result from the simulations, may easily be changed by altering the growth, recruitment or mortality parameters. If parameters for growth or recruitment are set to give a higher growth or recruitment, then the mean values of the stock, and the catch, will get higher in the simulations. In the reference simulation the growth parameters are set rather low, and the parameters in the recruitment function are estimated from data for the period 1946-1990.

The management strategies for cod and herring are based on VPA. For herring, an independent stock estimate is available. If accepted as a reliable estimate, then both stock size and total mortality is known. The actual fishing mortality is then computed from the reported catches, and natural mortality is computed as the difference between total mortality and fishing mortality. If no independent stock estimate is available, then the fishing mortality will be estimated using survey data (tuning indices), assumptions about natural mortality and a tuning algorithm. In the scenario model, cod management is based on VPA with Laurec-Shepherd tuning.

The two ways to perform the VPA computations will be referred to as simulations with and without tuning in the management procedure. In our scenario model the cod management is always with tuning, and herring management is without tuning. In simulations with tuning the tuning indices are generated from the stock size (see Fishery and research surveys).

Simulations with and without tuning will differ when studying special factors, for example when analyzing the effect of an extra fishing mortality. When the stock size is assumed know, the total mortality is also assumed known every year. If total fishing mortality is higher than the reported catch, the extra fishing mortality is added to the natural mortality. The resulting high natural mortality is used when computing the quota. In simulations where a tuning algorithm is used to determine the fishing mortality, the natural mortality is assumed known. If there is an extra fishing mortality which is not reported, the fishing mortality rate (F-value) will be kept, but the stock estimate will be adjusted to match the catch. It is important to distinguish between the two methods when discussing management procedures for cod and herring.

All summary statistics presented in this section are mean values from 100 years in one single simulation.

Simulation with F_{high}, F_{med} and F_{low}

These strategies are alternative management strategies for cod. For herring they are, however, not that interesting, because of the nature of herring recruitment. In the simulations with these cod strategies, the simulation model is parameterized so as to give low growth. The recruitment parameters in the Beverton-Holt function are estimated using the data for 1946-1990.

In the simulations it is assumed that recruitment does not increase further when the spawning stock biomass (SSB) exceeds 1 mill.tonnes. When computing the ratio between the number of recruits and the spawning stock, the size of the spawning

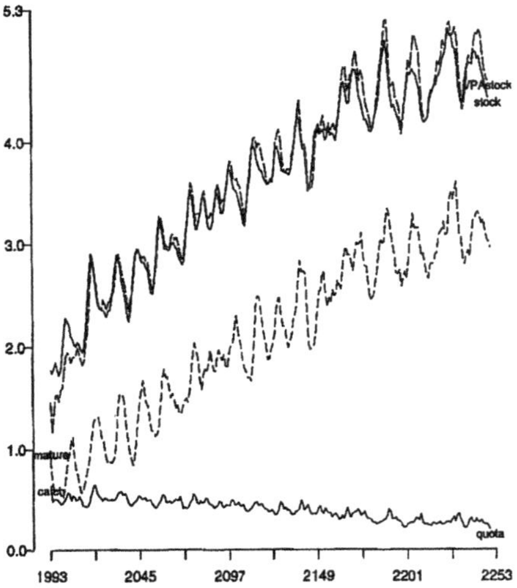

Figure 3. Simulation for cod with strategy T_{med}, and without SSB limit.

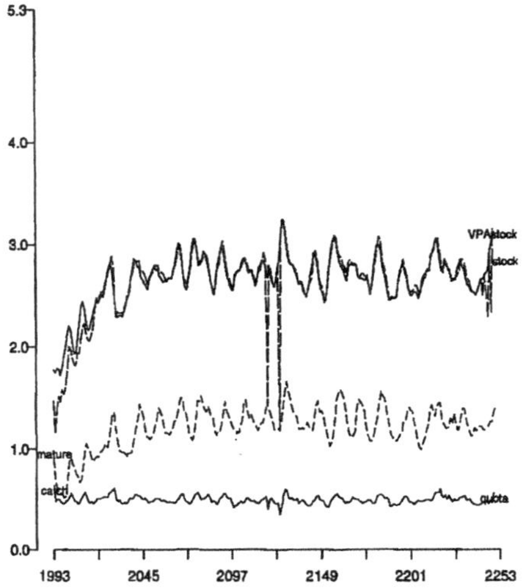

Figure 4. Simulation for cod with strategy T_{med}, and SSB limit of 1 mill. tonnes.

stock is nominally reduced to 1 mill.tonnes, when exceeding this limit. This is done because we do not want low quota values with a large spawning stock. When the spawning stock biomass increases, the estimated F-value for the quota decreases. This includes that even if the stock increases, the quota will not increase, and it might even decrease.

Figure 3 shows a simulation with F_{med}, where there has been no nominal limitation in the spawning stock when computing the ratio number and the quota. The figure shows that the stock grows slowly, but steadily during the simulation period, and the growth continues until the quota is nearly zero, and the stock is very large. This simulation has been run for about 250 years, and it illustrates that the stock grows very slowly. In this simulation the tuning indices are generated without noise.

F_{med} should determine a quota which should preserve the stock on a constant level. An extra fishing mortality of about 10% will stabilize the stock soon. Figure 4 shows a simulation with F_{med}, but with SSB nominally set at 1 mill.tonnes when the limit is exceeded. The stock stabilizes on a stock size of about 2.7 mill.tonnes and a quota of 0.49 mill.tonnes. The tuning indices are still generated without noise. The lower dashed curve shows the size of the mature stock, and the lower solid curve shows the quota which is equal to the catch. In the sequel the SSB limit is imposed.

Figure 5 shows a simulation with F_{high}. The fluctuations in stock size during the simulation period is large, and several times the mature stock has nearly disappeared. When the stock is depleted, the quotas get very low, and the stock recovers. After the stock having recovered, the quotas are increased, so that the mature stock once more almost disappears. These fluctuations tell that the F_{high} strategy tends to give too high quotas.

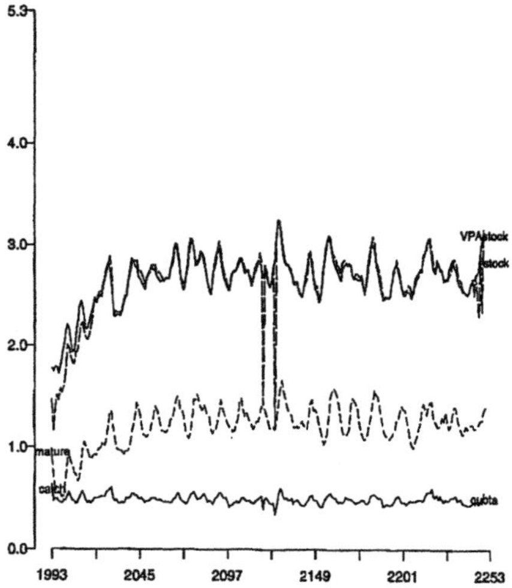

Figure 5. Simulation with management strategy F_{high} for cod.

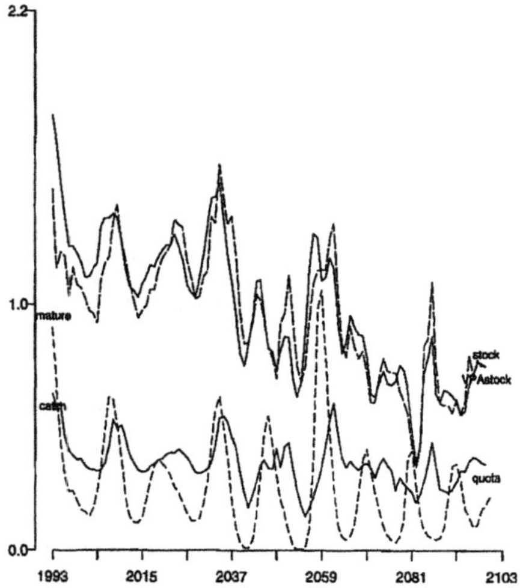

Figure 6. Simulation with management strategy F_{low} for cod.

In Figure 6 a simulation with F_{low} is shown. The stock grows until it stabilizes on a level about 3.3 mill.tonnes. This level is higher than the corresponding level when F_{med} is used as management strategy, but the quotas are lower. The stabilization of the stock is due to the nominal limitation of SSB in the quota computations. Without this limitation, the stock would have grown during the whole simulation period, and the growth would have been faster than with the F_{med} strategy (see Figure 3).

In the simulations presented in Figures 5 and 6, the tuning indices are generated with noise. However, as seen from the curves comparing the model stock (upper solid curve) and the VPA-estimated stock (upper dashed curve), there is more noise in the indices in the F_{low}-simulation than in the F_{high}-simulation.

Simulations with F_{max} and F_p

These management strategies are alternatives for both cod and herring. Management of herring based on these strategies are treated in the sections Management of herring and Management of herring with overfishing, and in this section we will discuss the cod management. Figure 7 shows a simulation with F_{max} strategy, and we see that both the stock size and the catch stabilize. The mean stock size is 2.9 mill.tonnes, and the mean catch is 0.49 mill.tonnes. This is probably close to the optimal stock size.
A simulation with the strategy $F_{0.1}$ gives a higher stock size than F_{max} (see Table 6). The reason for this is that $F_{0.1}$ always will determine a lower quota than F_{max}, for a given stock size, and thus results in a larger stock. The higher stock size does not compensate the lower quotas, though, and the mean catch during the simulation period is lower than with the F_{max}-strategy.
The results from simulations with F_{max} and F_{med} are quite similar. The mean catches

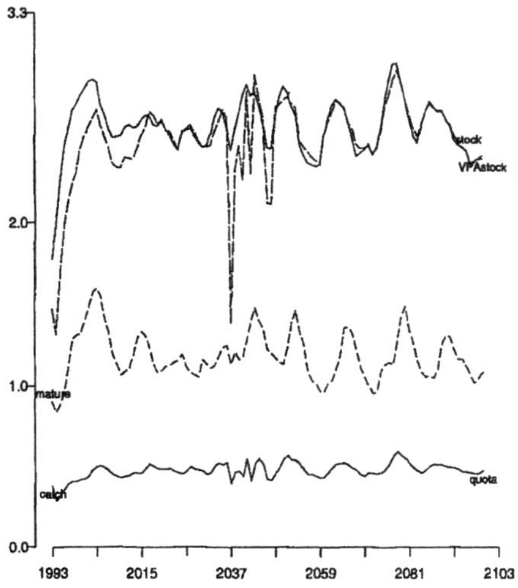

Figure 7. Simulation with strategy F_{max} for cod.

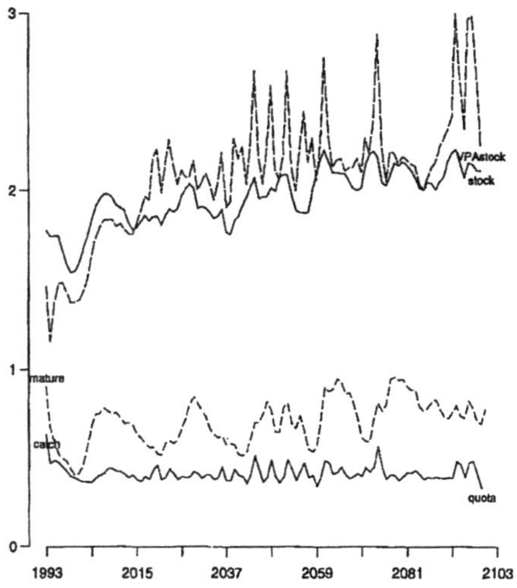

Figure 8. Simulation with strategy F_{med}, and with low growth, cod.

are approximately identical, but the stock size is slightly higher and the fluctuation in the estimated F-value is less with the F_{max}-strategy compared to the F_{med}-strategy.

The tuning routine

Using a tuning routine to estimate the fishing mortality from survey data, increases the fluctuations in the quota. The reason for this is an increased error in the estimated fishing mortalities, which give a quota based on incorrect assumptions. In our scenario model, the VPA analysis for cod is performed with Laurec-Shepherd tuning (Laurec & Shepherd, 1983).

The reference simulation with F_{med} is shown in Figure 4, and in this simulation there is no noise in the survey indices. If the indices are generated with noise, the VPA-estimate fluctuates around the value of the model stock, as illustrated in Figure 6.

The simulation shown in Figure 7 is done without noise in the survey indices. The

Table 6. Cod management. Stock, Mature and Catch are biomass in unit mill. tonnes, while Recruit are the number of recruits in unit billion.

Strategy	Stock	Mature	Catch	Recruit	F-value
F_{max}	2.9	1.19	0.49	0.76	0.24
$F_{0.1}$	3.5	1.80	0.43	0.74	0.16
F_{med}	2.7	1.07	0.49	0.77	0.27
F_{low}	3.3	1.61	0.44	0.72	0.18

difference between the model stock and the VPA estimate is small, except for one single year. The reason for this large deviation is that the tuning estimate for the fishing mortality this year is far from correct. Figure 4 gives another example.

Figure 8 shows a simulation with F_{med}-strategy, and with low individual growth. In this simulation the difference between the model stock and the VPA estimate is larger. The stock is generally overestimated, and there are several years when the deviation in the VPA estimate is substantial. Simulations with extremely low fishing mortality, may also give VPA estimates with substantial deviations.

When the survey indices are correct, the Laurec-Shepherd tuning routine normally results in a good estimate of the fishing mortality, provided that there is normal growth and the fishing mortality is not too small. There may, however, be single years with large deviations in the estimate of the fishing mortality. In years with low growth the tuning routine will underestimate the fishing mortality. If the survey indices are noisy, the estimated fishing mortalities will fluctuate around the true values.

Natural mortality rate

Mortality rate in the simulation model

In the simulations discussed above, the management strategy F_{max} was found to be slightly better than F_{med}. The real natural mortality rate was M=0.2. In this section we will study the behaviour of the strategies when this mortality rate varies, still keeping the nominal mortality rate in the VPA-computations at M=0.2.

Varying the M-value in the model over the interval [0.1, 0.3], the optimal mean catch also will vary. Table 7 shows optimal catch values when the simulation conditions are identical, except for this natural mortality rate.

The management strategies F_{max} and F_{med} both depend on the natural mortality rate, M, the dependence is, however, different. The F-value computed by F_{max} increases when the natural mortality increases, while the F-value set by F_{med} decreases, see the Natural mortality dependence section. When the VPA-value of M is kept fixed at M=0.2, it is, however, minor difference in F-values for both strategies, as seen from Table 8. In the reference simulation with M_{true}=M_{VPA}=0.2 the results of F_{med} and F_{max} are about identical, (see Simulations with F_{max} and F_p). When M_{true} is lower than M_{VPA}=0.2, F_{max} seems to be the best of the two strategies, and when M_{true} is higher than M_{VPA}, the two strategies seem to perform quite equally, even though F_{max} tends to give slightly higher stock values.

Table 7. Optimal catches when the nominal natural mortality rate in the VPA computations for cod is set to M=0.2, and with varying real natural mortality.

Natural mortality (M)	Optimal catch (mill.tonnes)
0.1	0.7
0.2	0.5
0.3	0.3

Nominal mortality rate in the VPA routine

The true natural mortality rate is now kept fixed (M=0.2), while the nominal mortality rate in the VPA-computations is varying. The state of nature is thus fixed, and we will see how the management strategies behave when the nominal mortality rate used in the VPA computations is not correct.

Simulations have been run for the strategies F_{max} and F_{med}, with the M-values 0.1, 0.2 and 0.3 in the VPA computations. The results are shown in Table 9. We see that the F_{med} strategy gives approximately identical values for stock and catch for all the three M-values. The incorrect stock estimates in the VPA are compensated by choice of F-values in the opposite direction.

Cod management with overfishing

In cod management the size of the stock is unknown. The fishing mortality-rate, the F-value, is computed from the tuning algorithm, based on tuning indices, and then stock size is computed from the reported catch, natural mortality and fishing mortality. The reported catch is usually identical to the quota, but the quota may be overfished. Figure 9 shows a simulation with 40% overfishing. In this figure, percentage overfishing is the same for all length groups. The mean value of the catch and the model stock size is approximately identical to a corresponding simulation with no overfishing. The reason for this is that the VPA routine interprets the overfishing as an increased natural mortality, and the stock estimate is therefore set too low. Because

Table 8. Varying natural mortality for cod. Real natural mortality rate in the simulation model is given by M=0.1 or M=0.3.

Strategy	Stock	Mature	Catch	Recruit	F-value	M	VPA-M
F_{med}	2.2	0.77	0.28	0.80	0.30	0.3	0.2
F_{max}	2.5	0.98	0.24	0.84	0.24	0.3	0.2
F_{med}	2.8	1.01	0.78	0.70	0.31	0.1	0.2
F_{max}	3.2	1.39	0.80	0.69	0.24	0.1	0.2

Table 9. Management by different M values in the VPA computations for cod.

Strategy	Stock	Mature	Catch	Recruit	F-value	VPA-M
F_{max}	4.2	2.42	0.33	0.73	0.19	0.1
F_{max}	2.9	1.20	0.49	0.76	0.24	0.2
F_{max}	1.8	0.43	0.47	0.69	0.36	0.3
F_{med}	2.7	1.06	0.47	0.76	0.36	0.1
F_{med}	2.7	1.07	0.49	0.77	0.27	0.2
F_{med}	2.4	0.84	0.51	0.77	0.23	0.3

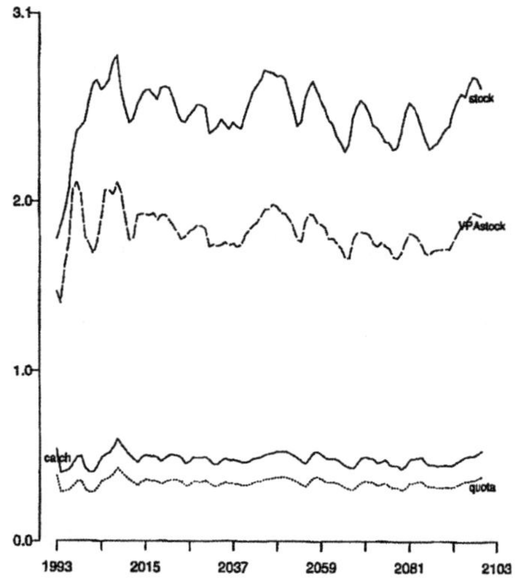

Figure 9. Simulation with strategy F_{max} for cod, with 40% overfishing.

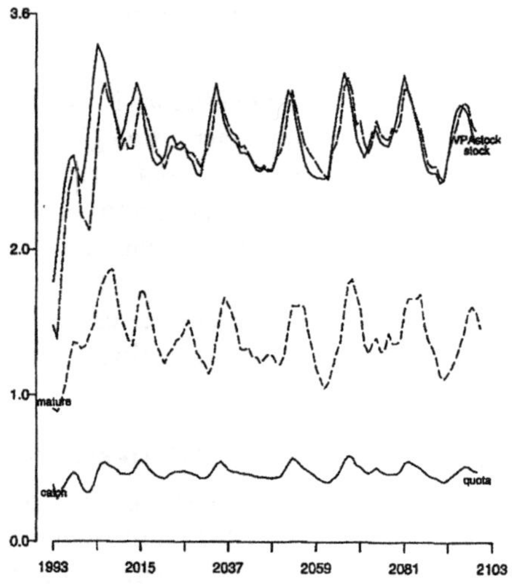

Figure 10. Simulation with strategy F_{max} for cod, and lower limit 35 cm.

of the lower stock estimate the quota is also lowered, and the result is a situation where the total catch (quota plus overfished quantity) is approximately identical to the quota in a corresponding simulation with no overfishing. This is also illustrated in the uncertainty experiments, see the Mean yearly catch section.

With a sudden strong increase in the underreporting, the stock size will decrease for some years before the management procedure starts to reduce the quota.

Table 10 gives simulation results for the strategies F_{max} and F_{med}. The mean stock size and total catch seem to be independent of whether or not the quota is overfished. The overfishing mainly results in reduced quotas (reported catches) because of the low VPA-estimates, as illustrated in Figure 9.

Different selection patterns

In the simulations presented in the preceding sections, no catch is performed on cod shorter than 45 cm. The length groups 45-55 cm have fishing mortality increasing with

Table 10. Cod management with overfishing, stock and catches in unit mill. tonnes.

Strategy	Stock	VPA stock	Catch	Quota	F-value	Overfishing
F_{max}	2.9	1.19	0.49	0.49	0.24	0
F_{max}	2.8	1.12	0.48	0.34	0.24	40%
F_{med}	2.7	1.07	0.49	0.49	0.27	0
F_{med}	2.7	1.10	0.47	0.34	0.27	40%

length, while cod longer than 55 cm have constant fishing mortality. This pattern defines a selection pattern.

Some simulations have been run with the lowest catchable length set to 35 cm, and Table 11 displays the results. The mean catch gets lower (0.02-0.03 mill.tonnes) when the length limit is reduced with 10 cm, but the mean stock size is approximately the same. The mature biomass is larger and the immature smaller. The main reason for the lower catch is that the changed selection pattern gives lower F-values when the quota is computed. The higher fishing mortality on the young cod is compensated by a larger mature stock and thus a higher recruitment.

Reducing the lowest catchable length from 45 cm to 35 cm reduces the mean catch by some 5%. Reducing the lowest allowable length further will also reduce the catches further.

Management of herring

Herring is managed by the F_p strategy, and in the probing simulations the values of p are 0.1 and 0.3. The herring recruitment has large fluctuations, and approximately every tenth year the recruitment is extremely high. The number of normal years between two years of high recruitment is determined stochastically, as described earlier in the Recruitment section. In this way the number of years with high herring recruitment will vary between simulations, and the resulting mean stock and mean catch will also vary.

In Table 12 some simulation results are shown. The simulations presented in this table are from single simulations, and with average results. Due to the stochastic recruitment, results from single simulations may vary substantially, as explained above. The strategy $F_{0.1}$ seems to give a higher catch than $F_{0.3}$, but the stock size is much lower when the strategies are applied without any restrictions.

When using the strategy $F_{0.1}$ the catch varies from 0.20 to 1.26 mill.tonnes, and this reflects the recruitment variation. To reduce the yearly catch fluctuations, the strategy could be modified by introducing an upper catch limit. An upper catch limit of 0.6 mill.tonnes has positive effects for the $F_{0.1}$-strategy, the mean stock size increases while the mean catch is unaffected. An alternative modification is to define a lower limit for the spawning stock biomass, and to have only a minimal catch when the spawning stock gets below this limit. A spawning stock limit of 2.5 mill.tonnes results in an increase in mean stock size, but the mean catch decreases as a consequence of

Table 11. Varying lowest allowable catch limit in cod management, stock and catches in unit mill. tonnes.

Strategy	Stock	Mature	Catch	Recruit	F-value	Limit
F_{max}	3.0	1.40	0.47	0.80	0.19	35 cm
F_{max}	2.9	1.19	0.49	0.76	0.24	45 cm
F_{med}	2.7	1.14	0.46	0.80	0.22	35 cm
F_{med}	2.7	1.07	0.49	0.77	0.27	45 cm

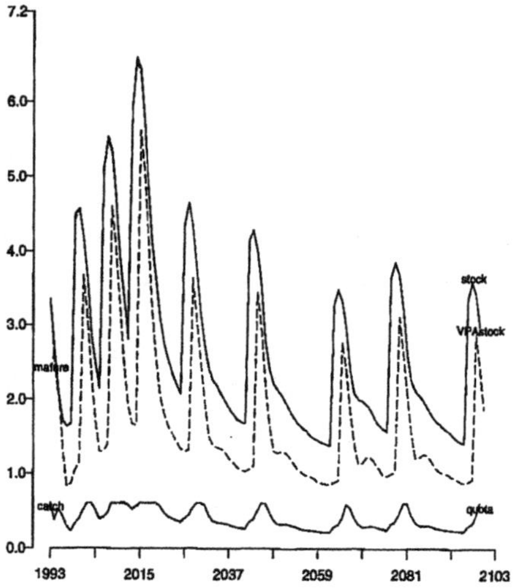

Figure 11. Strategy $F_{0.1}$ for herring, upper catch limit 0.6 mill. tonnes.

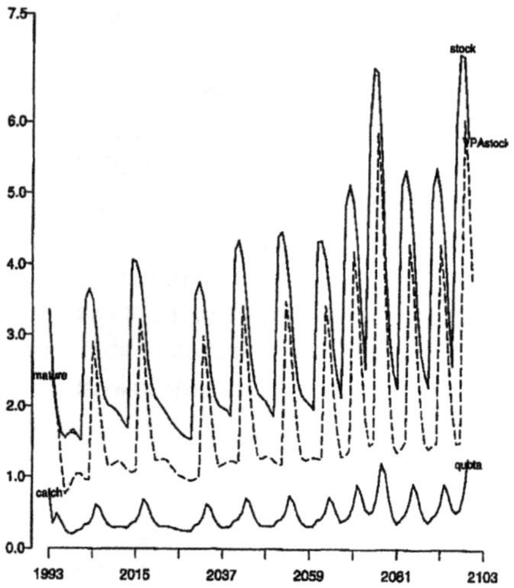

Figure 12. Strategy $F_{0.1}$ for herring, no upper catch limit.

many years with almost no catches. Simulation results for both modifications are shown in Table 12.
The herring simulations seem to indicate that the best strategy is $F_{0.1}$, and with an upper catch limit. An example of a simulation with this strategy is shown in Figure 11. Figure 12 shows a simulation with $F_{0.1}$, but without catch limit. The figure shows that the quota is more variable than in the simulation with upper catch limit.

Management of herring with overfishing

In the herring management the correct stock size and the total mortality are assumed known. The fishing mortality (F-value) is computed from the reported catch, and no tuning routine is involved. Mortality that is not explained by the fishing mortality, is

Table 12. Herring management (all numbers in mill.tonnes).

Strategy	Upper catch limit	Spawning stock limit	Total stock	Mature stock	Catch Mean	Catch Min	Catch Max
$F_{0.1}$	None	None	4.2	1.9	0.48	0.20	1.26
$F_{0.3}$	None	None	6.2	3.4	0.44	0.17	0.83
$F_{0.1}$	0.6	None	4.7	2.3	0.50	0.27	0.60
$F_{0.3}$	0.6	None	6.0	3.3	0.42	0.17	0.60
$F_{0.1}$	0.6	2.5	6.4	3.6	0.41	0.04	0.60
$F_{0.3}$	0.6	2.5	6.2	3.5	0.36	0.04	0.60

then interpreted as natural mortality. Fishing mortality which is not reported, for example because of overfishing, will therefore be included in the natural mortality. Herring management in a situation with overfishing, is then biologically equivalent to herring management in a situation with higher natural mortality and no overfishing.

Table 13 shows results from simulations with 40% overfishing of the quotas, which means the total catch is always 40% higher than the quota. The simulations show that the total catch is approximately the same, irrespective of overfishing or not, but in a situation with overfishing the stock size is considerably lower. An upper quota limit of 0.43 mill.tonnes in a situation with 40% overfishing has similar effect to the limit of 0.6 mill.tonnes in the situation with no overfishing.

Management of capelin

The capelin procedure is not based on VPA computations, but uses the CapTool method (Bogstad *et al.*, 1992), and quota is only set for the mature stock in the winter fishery. Since the mature stock dies after spawning, the fishery has only effect on the recruitment and the predation in other species. An autumn fishery on the immature stock will have greater dynamic impact on capelin. The management target is to maintain the spawning stock at a given level (spawning target), in most of the simulations 0.5 mill.tonnes. If the predicted spawning stock is lower than the target value, the CapTool quota is zero.

Table 13. Herring management with 40 % overfishing (unit is mill. tonnes).

Strategy	Upper quota limit	Total stock	Mature stock	Catch Total	Catch Quota	Catch Min	Catch Max
$F_{0.1}$	None	3.3	1.3	0.50	0.36	0.18	1.28
$F_{0.3}$	None	4.7	2.3	0.47	0.34	0.23	1.11
$F_{0.1}$	0.6	3.2	1.3	0.47	0.34	0.14	0.85
$F_{0.3}$	0.6	4.7	2.3	0.46	0.33	0.19	0.86
$F_{0.1}$	0.43	3.7	1.7	0.46	0.33	0.16	0.62
$F_{0.3}$	0.43	5.5	2.8	0.49	0.35	0.19	0.62

Table 14. Capelin management, mean values (mill.tonnes) over 100 years.

Mature stock limit	Minimum quota	Stock	Spawning biomass	Catch	Years with low catch
0.5	None	3.2	0.43	1.0	40
0.5	0.1	2.9	0.35	1.0	40
0.1	None	2.8	0.10	1.0	50

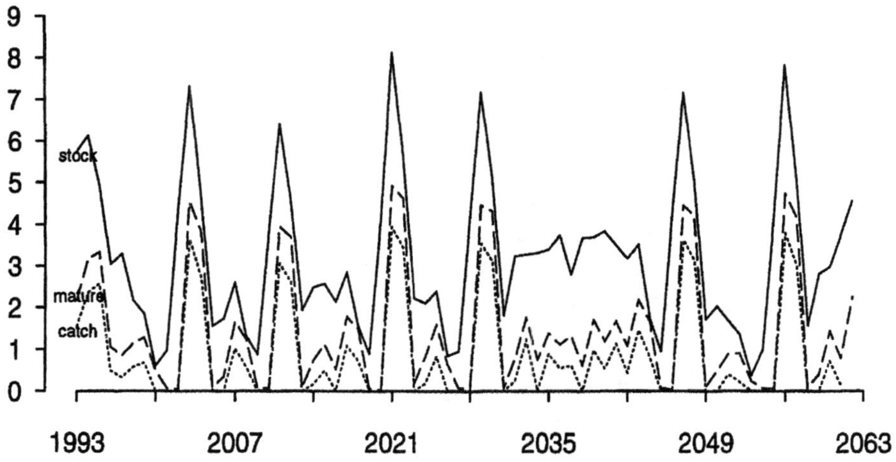

Figure 13. Capelin, CapTool management.

The effect of reducing the spawning stock target is rather small. Introducing a minimum quota, which is always caught, results in increased catches when the mature stock is on a low level. However, this will also reduce recruitment, and the periods with low stock values will last longer. The result is that the mean stock size is reduced with 0.3 mill.tonnes, and the mean catch is approximately identical to the result with no minimum quota. Reducing the spawning stock target to 0.1 mill.tonnes, has similar effects to the minimum quota. In both cases, the mean quota is higher than the resulting mean catch. The reason is that the mean quota several years is set higher than the total spawning stock, and so it is not possible to reach the quota.

From the above we can conclude that the effect of varying the management criteria seems small. The main reason is weak dependence between the spawning stock biomass and recruitment. The halfvalue in the Beverton-Holt function (3) is supposed to be low, and a small spawning stock may well give high recruitment when the environments are favourable. Figure 13 shows the fluctuations in the stock and the catch during a simulation with mature stock limit 0.5 mill. tonnes, and no minimum quota.

UNCERTAINTY SCENARIOS

In addition to the probing scenarios, a set of uncertainty scenarios are simulated. Such scenarios should partly be based on results from probing scenarios, but more importantly, they should be laid out in co-operation with experts in fisheries biology, economics and management. The uncertainty scenarios aim at spanning a plausible region in the state space. The variation in performance measures over scenarios will be informative with respect to issues of robustness (the likelihood of severely depleting the ecosystem) and efficiency (the mean performance in scenarios central in the plausibility region). The experiment presented in this article is mainly meant as an illustration, and was not designed by an expert group.

Experimental factors

In this uncertainty experiment we will differ between two kinds of experimental factors:

1. Biological factors, like recruitment.
2. "Human" factors with consequences in management, like overfishing.

From the biological group we have chosen only to include recruitment-related factors. From the "human" group we have chosen both fishery-related factors, and plain management-related factors. This group is named "human", because the factors include human decisions in one way or another.

In addition to the experimental factors, also the management strategies are varied. The total management strategy is composed of three single-species strategies, and similarly to the experimental factors, levels are defined for each. Also the management strategies will sometimes be referred to as "factors", even if they are not experimental factors at the same level as the biological or the human factors.

The experimental factors are defined binary, while the management strategies not necessarily are so. In this section both experimental factors and management strategies are described, and in the next section the whole experimental design is outlined.

Biological factors

As mentioned above, only recruitment-related factors are included. Recruitment is modeled according to Beverton-Holt-functions subjected to stochastic noise, see equation (3). Recruitment level is given by the expected value, and is varied through the parameters *MAXR* and *H*. Variation in recruitment is due to the white noise, and is varied through the standard deviation in the corresponding noise distribution. We have chosen 5 recruitment-related factors, one capelin-factor (recruitment variation) and two herring- and cod-factors (both level and variation), and they are summarized in Table 15. The cod variations (VCOD) do need an explanation. From the tabulated values we observe that the standard deviation is larger than the maximum level, which means that there is a high probability of drawing a negative number of recruits. This is of course not allowed, both lower and upper limits for number of recruits are defined as percentages of the expectation. If the simulated number exceeds those limits, the proper limit value is used. This parameterization should, perhaps, be improved.

"Human" factors

With the name "human" factors, we mean all fisheries and research factors that are not biological. In the present simulation experiment we have only defined "human" factors that directly involves the cod stock: factors related to fishery, research surveys and VPA. Table 16 presents a summary of these factors, and a closer description is given below.

VPA

When running the VPA-program to estimate the cod stock, a nominal natural mortality rate, *M*, is used as a known constant. In our simulations the true natural mortality rate

Table 15. Table summarizing the biological factors. *MAXR* and *H* are the parameters in the Beverton-Holt-function (3), while $\sigma_{species}$ is the standard deviation in the distribution giving the stochastic variation in recruitment. The numbers in parenthesis are the values in years of extreme recruitment.

Factor	Abbr.	Reference level	Alternative level
Level of recruitment cod	RCOD	*MAXR*=1.02(1.74) bill. *H*=0.18 mill.t	*MAXR*=0.77(1.31) bill. *H*=0.14 mill.t
Level of recruitment herring	RHER	*MAXR*=26.52(280.63) bill. *H*=1.29 mill.t	*MAXR*=19.89(210.47) bill. *H*=0.97 mill.t
Variation in recruitment cod	VCOD	σ_{cod} = 1.75 bill.	σ_{cod} = 3.5 bill.
Variation in recruitment herring	VHER	σ_{her} = 3.0 bill.	σ_{her} = 6.0 bill.
Variation in recruitment capelin	VCAP	σ_{cap} = 200 bill.	σ_{cap} = 400 bill.

Table 16. A summary of the "human" experimental factors. All factors include only the cod stock (fishery).

Factor	Abbr.	Reference level	Alternative level
Natural mortality rate (in VPA)	M	*M* = 0.2 (true value)	*M* = 0.1 (underestimating)
Survey indices, noise	I	*k* = 0 (no noise)	*k* = 1
Overfishing	O	No overfishing	Norwegian and Russian 50% Other nations 200%
Discards	U	No discards	Discard of cod less than 47 cm
Age reporting of catches	A	Correct	Young fish under-reported Old fish over-reported

is set to M=0.2. In the scenarios the nominal VPA-value of M is treated as an experimental factor. The reference level will be the ideal situation where the VPA-constant matches the real mortality, and at the alternative level the nominal value used in VPA is lower than the natural mortality.

Tuning indices

During the year both Norway and Russia perform research surveys in the Barents Sea, and data from these surveys serve (among other things) as input to the VPA. We have modeled 5 sets of survey data (tuning indices), used for estimating the cod stock at different points of time during the year. The data are indices for the age groups 3-7. In the simulations the total index for each survey is proportional to the total cod stock, but subjected to some noise:

$$Index = a \cdot B + noise \ ,$$

with B being the total stock number in unit 1000. The function parameter a is specific for each survey, and so is the standard deviation in the noise distribution. The age distribution is, however, correct in all simulations. The standard deviation is modeled

$$\sigma_{survey} = k \cdot s_{survey} \ ,$$

with k being a common factor to all the distributions. We have defined this k-value to be an experimental factor, with reference level k=0. At this level the survey indices are generated without noise. As the alternative level, the value k=1 is chosen. The following parameter values are a result from linear regression on VPA-data and historical tuning indices (ICES, 1993), and are used in the simulations (unit of standard deviation is 1000):

Survey:	1	2	3	4	5
a	0.00015	0.00025	0.00010	0.00020	0.00050
s_{survey}	50	200	50	25	300

Fishery

The cod fishery is divided into 6 independent fisheries (fleets). First the total quota is distributed to 3 national groups: Norway (45%), Russia (45%) and other nations (10%). Then the Norwegian quota is distributed to 4 groups of fishing gears: trawl (27%), net (37%), hand line (18%) and seine (18%). The length distributed selection patterns are roughly computed from observed age distributed patterns and age-length distributions. Both the distribution of quotas and the fleet specific selection patterns are constant through all simulations, while the amount of overfishing and discards are defined as experimental factors, together with the age reporting of the catches. The amount of overfishing is set to a percentage of the quota. With discard is meant that all fish under a lower length are dumped, and sufficient catch is taken to fill the quota with larger fish. When there is discard, this strategy is followed throughout the cod fishery. If there also is overfishing, the discard strategy is followed for the overfished quantity as well. The age reporting has consequences primarily for the VPA-estimation of the cod stock, where catch-by-age data are the main input. At reference level there is correct age reporting, but at the alternative level the old age groups are overreported (-estimated) at the expense of the young ones. The following percentages are used when age-reporting is on alternative level:

Age	3	4	5	6	7	8-12	13+
Reported %	9.5	1.7	1.7	45.5	105	110	100

This means that when reporting catch-by-age, the true catches are multiplied with the given percentage before reporting. If, as an example, the true value is 100 tonnes of 6-year olds, the reported value is set to $100 \cdot 0.455 = 45.5$ tonnes.

Management strategies

In the present experiment the total management strategy is a combination of 3 single-species strategies, and we define a factor for each single species strategy. The capelin strategy is based on a spawning target, and the factor has 2 levels. Both the cod and the herring strategy is based on VPA-estimates for the stock, and these factors have 3 and 5 levels, respectively. The herring strategy is either a constant nominal fishing mortality rate or based on the yield per recruit function. The cod strategy might in addition to those two, be based on the spawning biomass per recruit function. Both strategies based on the yield per recruit function and strategies based on spawning biomass per recruit are updated yearly during the simulations. Table 17 gives a summary of all three management factors.

Experimental design

The 13 factors described above are combined in the system of uncertainty simulations. Since the number of combinations is too high, combinatorial methods of experimental design (Box *et al.*, 1978) are used to find a system of balanced combinations of manageable size. When defining the experimental design, we will make a distinction between the management strategies on the one hand, and the experimental factors on

Table 17. A summary of the management strategies. The capelin strategy gives the spawning stock target, while the herring and cod strategies give the fishing mortality rate F.

	Management strategy		
	Capelin	Herring	Cod
Reference level	500 000 tonnes	$F_{0.3}$	F_{med}
Alternative 1	100 000 tonnes	F=0.12	F_{high}
Alternative 2	—	F=0.20	F_{low}
Alternative 3	—	—	F_{max}
Alternative 4	—	—	F=0.26

the other. The last group consists of 10 binary factors. These are combined as follows:

1. One scenario with all factors on reference level.
2. 10 scenarios with exactly one factor on alternative level.
3. A 2^{10-5} fractional factorial design.

This gives a total of 43 different uncertainty scenarios. For the management strategies there are only 30 possible combinations. A random sample of 13 of these are given multiplicity 2, to get a total of 43. The uncertainty scenario with all factors on reference level is combined with the management strategy with all single species strategies on reference level, while the rest are combined at random. The resulting design is given in Table 18. All scenarios are simulated with 3 replicates.

Performance measures

The performance measures should summarize the development of the ecosystem and the fishery, given the scenario and the management strategy. The specification of performance measures should be done in co-operation with experts in biology, resource economy and fisheries management. This specification is non-trivial. The performance measures to be considered below might appear a bit naive. They are presented to illustrate the method, and to provoke further work on specifying better performance measures.

Catches

The catches give rise to several possible performance measures. The most obvious ones are mean yearly landed catches over the response period, which of course are wanted as high as possible. In principle, it is the total landings that describes the management strategy, but since the three species do have unequal (economic) value it would not be correct just to add the weights. One could introduce economic value of the catches, and add these to get an overall total catch response. However, since economic considerations are not explicitly included in the simulation model, we will not introduce such a variable in the present analysis.

Mature biomass

The purpose of this response is to describe how the state of the various fish stocks varies with the strategies. It is not obvious how to measure the "state of the stock" over the response period. Our expert group agreed to define lower limits for safe stock biomasses, and we count the number of years when the stocks are under these limits. The lower limits were defined for the mature stock:

Cod:	lower limit 0.2 mill.tonnes
Herring:	lower limit 1.0 mill.tonnes (alternatively 2.5 mill.tonnes)
Capelin:	No lower limit, due to the enormous natural fluctuations in capelin stock.

For herring we will use an intermediate limit in the present analysis, namely a lower limit on 1.75 mill.tonnes. We thus define two response variables, describing the mature

Table 18. A table summarizing all scenarios to simulate. 0 means that the factor is on reference level, while a positive integer gives the alternative level.

	RCOD	RHER	VCOD	VHER	VCAP	M	O	U	I	A	SCAP	SHER	SCOD
1	0	0	0	0	0	0	0	0	0	0	0	0	0
2	1	0	0	0	0	0	0	0	0	0	0	0	4
3	0	1	0	0	0	0	0	0	0	0	0	2	3
4	0	0	1	0	0	0	0	0	0	0	1	0	4
5	0	0	0	1	0	0	0	0	0	0	0	1	4
6	0	0	0	0	1	0	0	0	0	0	1	2	3
7	0	0	0	0	0	1	0	0	0	0	1	1	4
8	0	0	0	0	0	0	1	0	0	0	0	2	4
9	0	0	0	0	0	0	0	1	0	0	1	2	4
10	0	0	0	0	0	0	0	0	1	0	1	1	0
11	0	0	0	0	0	0	0	0	0	1	0	2	0
12	0	0	0	0	0	1	1	1	1	1	1	0	1
13	0	0	0	0	1	1	1	1	0	0	1	2	0
14	0	0	0	1	0	1	1	0	1	0	0	1	1
15	0	0	0	1	1	1	1	0	0	1	0	2	1
16	0	0	1	0	0	1	0	1	1	1	1	0	2
17	0	0	1	0	1	1	0	1	0	0	0	1	2
18	0	0	1	1	0	1	0	0	1	0	0	0	3
19	0	0	1	1	1	1	0	0	0	1	0	2	2
20	0	1	0	0	0	0	1	1	1	1	0	1	4
21	0	1	0	0	1	0	1	1	0	0	1	1	4
22	0	1	0	1	0	0	1	0	1	0	0	2	4
23	0	1	0	1	1	0	1	0	0	1	1	0	0
24	0	1	1	0	0	0	0	1	1	1	0	1	0
25	0	1	1	0	1	0	0	1	0	0	0	0	1
26	0	1	1	1	0	0	0	0	1	0	1	1	0
27	0	1	1	1	1	0	0	0	0	1	0	2	0
28	1	0	0	0	0	0	0	0	0	1	1	0	1
29	1	0	0	0	1	0	0	0	1	0	1	2	0
30	1	0	0	1	0	0	0	1	0	0	0	1	1
31	1	0	0	1	1	0	0	1	1	1	1	1	1
32	1	0	1	0	0	0	1	0	0	1	0	0	2
33	1	0	1	0	1	0	1	0	1	0	0	2	1
34	1	0	1	1	0	0	1	1	0	0	1	0	2
35	1	0	1	1	1	0	1	1	1	1	1	2	1
36	1	1	0	0	0	1	0	0	0	1	0	1	2
37	1	1	0	0	1	1	0	0	1	0	1	1	2
38	1	1	0	1	0	1	0	1	0	0	0	0	3
39	1	1	0	1	1	1	0	1	1	1	0	2	2
40	1	1	1	0	0	1	1	0	0	1	1	0	3
41	1	1	1	0	1	1	1	0	1	0	1	2	2
42	1	1	1	1	0	1	1	1	0	0	0	1	3
43	1	1	1	1	1	1	1	1	1	1	1	1	3

stocks over the response period, the number of years with mature cod biomass less than 0.2 mill.tonnes and the number of years with mature herring biomass less than 1.75 mill.tonnes.

Analysis

In this section we will use the statistical computer package splus (Chambers & Hastic, 1992) to estimate hierarchical, linear models to describe the relation between the catch responses and the experimental factors. We use Splus notation to specify descriptive models. As an example, take

$$Y\tilde{}A*(B+C) \tag{6}$$

The interpretation is that the response Y depends on the factors A, B and C, and that there are interaction effects between A and B and between A and C. The * can be interpreted as a "multiplication" sign, and the formula is evaluated using the distributive law. Hierarchical models assure that when the interaction effect A* B is included, the main effects from both A and B are included as well.

Methods of interpretation

The most fundamental uncertainty related to fisheries management stems from the lack of knowledge, and imperfections and uncertainty associated with available knowledge. Understood variability and stochastic uncertainty can be explicitly accounted for in the management models, while imperfections and lack of knowledge can not easily be accounted for in the model itself.

Our approach to study the properties of a management strategy, particularly with respect to comparative efficiency and robustness, is by a simulation experiment of uncertainty scenarios. In this approach, the main factors of uncertainty are identified, and a plausible region in the state space is spanned by the levels of these factors. In the present experiment, there are K=10 binary factors. As an example, take one of these factors, VHER, representing the standard deviation in herring recruits in "normal" years at a given level of spawning stock, σ_{her}. Ideally, the level of this factor, and others, should have been determined by expert judgement. To illustrate, assume it is inferred that 3 bill.$\leq \sigma_{her} \leq$ 6 bill., with the two limiting values lying about equally far out to either side. One can think of the interval [3 bill.,6 bill.] as a symmetric confidence interval, in the sense that the judged probability of σ_{her}<3 bill. is about the same as the judged probability of σ_{her}>6 bill. The coverage probability of the interval is judged on the basis of available data and other less direct sources of information, by experts in fisheries biology. Unfortunately, these values have been chosen without proper input from experts, and that applies also to the levels of other factors in the present experiment. The results to be presented should therefore not be taken as definite results, but should rather be understood as illustrative of the method.

To further discuss methodology, assume that the main factors of uncertainty have been identified. Assume further that judgemental confidence intervals for the parameters of all the factors, preferably all with the same degree of confidence, have been determined. These span a convex, rectangular region of plausibility, or confidence. Under independence, the confidence of the true parameter vector being inside this box is $(1\text{-}\alpha)^K$, when $1\text{-}\alpha$ is the confidence of the individual intervals.

After having decided on an experimental plan, which often will be a small fraction of the 2^K factorial plan (with dichotomous uncertainty factors) partly crossed with a corresponding plan in the management strategies (see Table 18), a linear model is estimated for a given performance measure. This estimated model allows the mean response to be predicted for each point in the full 2^K factorial layout for each management strategy. In fitting the linear model, the ordinary statistical approach of including hierarchical terms spanned by significant interactions might be taken. This might lead to a reduction in the set of relevant multispecies management strategies to be considered. For the performance measure mean yearly catch of cod, it is, for example, found below that the response is not significantly dependent on how herring and capelin is managed. Thus, the relevant strategies are simply the 5 single species strategies for cod.

Having fitted the linear model, the predicted mean response is calculated for each relevant management strategy and for each of the points in the 2^K design. The distribution of these 2^K predicted response values does characterize the management strategy in question: The central tendency (median, mean) is a measure of efficiency; the (downward or upward) spread (standard deviation, median ± 5% point) measures the lack of robustness. One could also identify regions in parameter space with particularly good or bad performance, say exceeding the 10% or 90% quantile in the predicted response distribution.

Mean yearly catch

Cod

Due to the possibility of discards, only cod longer than 47 cm are included in the response variable. If the discard factor (U) is on alternative level, all fleets follow a strategy of discarding small fish if it is judged profitable. In our report files from the simulations it is not possible to differ between the discarded and the landed catches in the smallest length groups (cod shorter that 47 cm), and so we have chosen to disregard those groups from the response variable.

Discards and general overfishing of the quotas are included as experimental factors. There are thus two possible catch variables for cod: the total (landed) catches, and the reported catches. It is not obvious which one to choose as the appropriate "cod response". We will take a look at both, but with main emphasis on the total landings.

From Figure 14 the most important factors appear to be the management strategy for cod (SCOD), the recruitment variables for cod (RCOD and VCOD) and the nominal mortality in VPA computations (M). Starting with these, a linear model is sequentially expanded through a process of forward stepwise regression to gain a better fit. The final model constrained to be hierarchical and only involve two-factor interactions, is

$$Mean\ catch \sim SCOD*(M + RCOD + VCOD + I + VHER) + VCAP\,. \qquad (7)$$

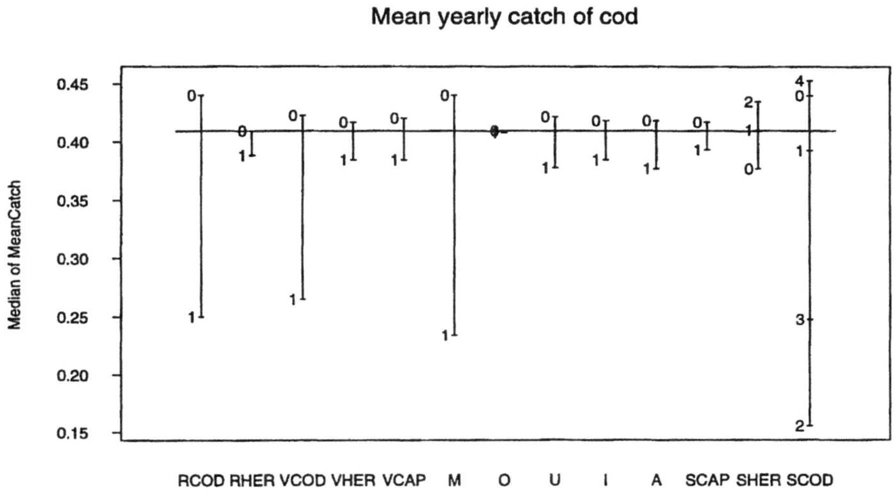

Figure 14. Marginal plot with total catch of cod as response variable. The vertical axis indicates median value when the experimental factors are on given level.

Thus the response variable "Mean catch" depends on the factors SCOD, M, RCOD, VCOD, I, VHER and VCAP. In addition to the main effects from these factors, there are two-factor interactions between the management strategy factor SCOD and all the other factors, except the capelin factor VCAP.

Neither the capelin strategy nor the herring strategy is significant for the cod catches. It is not very surprising that the herring strategy is not significant, since the herring present in Barents Sea is not subjected to fishery. That the capelin strategy is not significant does, however, seem a little surprising. The reason for this is the nature of the capelin strategy, CapTool. When computing the capelin quota, the cod consumption is taken into account, and the quotas are computed with boundaries set both by the spawning target (the capelin strategy), and by the cod consumption (independent of both cod and capelin strategy). The cod consumption is computed on the basis of the VPA-estimated cod stock.

The coefficient estimates for the model (7), describing the relation between the mean yearly cod catches over the response period and the experimental factors, are given in Table 19. The model fit is well, with a multiple R-squared equal to 0.99. Also plots of fitted values against the simulated data and residual plots, show that this model is very good in describing the cod catches.

The estimated model predicts a mean yearly catch of approximately 0.45 mill.tonnes when all experimental factors are on reference level (the intercept term). Reference level for the management strategy is F_{med}. The main effect from using strategy F_{high} instead (SCOD on level 1) is a lower predicted value, and as we can observe from the standard error, the reduction of approximately 0.05 mill.tonnes is significant (5%). Using strategy F_{max} or the constant strategy F=0.26 (SCOD on levels 3 or 4, respectively), does not give significantly different predictions from the F_{med}-strategy.

Table 19. Coefficient estimates, model (7), Unit is mill. tonnes.

RESPONSE: MEAN YEARLY CATCH OF COD

Main effects				
Factor	Level	Value	Std. Error	P-value
Intercept		0.452	0.005	0.000
SCOD	1	-0.045	0.012	0.000
	2	-0.276	0.021	0.000
	3	0.001	0.007	0.917
	4	0.004	0.006	0.514
M	1	-0.021	0.010	0.029
RCOD	1	-0.019	0.011	0.079
VCOD	1	0.021	0.008	0.007
I	1	-0.018	0.007	0.013
VHER	1	-0.013	0.007	0.075
VCAP	1	-0.008	0.004	0.051

Two-factor interactions				
SCOD-level	Second factor	Value	Std. Error	P-value
1	M	0.030	0.017	0.093
2		-0.058	0.015	0.000
3		-0.193	0.022	0.000
4		-0.049	0.013	0.000
1	RCOD	-0.001	0.016	0.941
2		0.067	0.014	0.000
3		0.004	0.015	0.802
4		-0.015	0.014	0.307
1	VCOD	-0.018	0.012	0.115
2		0.017	0.012	0.152
3		-0.003	0.013	0.807
4		-0.020	0.012	0.084
1	I	-0.001	0.010	0.958
2		0.027	0.010	0.007
3		0.023	0.014	0.105
4		0.012	0.010	0.228
1	VHER	0.032	0.009	0.001
2		0.020	0.009	0.030
3		0.016	0.013	0.237
4		0.015	0.010	0.134

However, the above conclusion depends on the assumption that all other factors included in the model remains on reference level. Taking a closer look on the coefficient estimates, we find that when the nominal VPA-mortality factor (M) is on reference level (matches the real mortality exactly), F_{med} and the constant F-strategy (SCOD on level 0 and 4, respectively) perform quite similarly. When the M-factor is on alternative level, the constant F strategy gives lower predictions than the F_{med} strategy. An explanation for this is that when the VPA-mortality is correct, F_{med} will be close to the chosen F-value in the constant F-strategy. If the VPA-mortality is not correct, however, then the F_{med} strategy seems to compensate for the error using the tuning indices, while the constant F is not able to do so. Also the F_{max}-strategy performs quite similarly to the F_{med} when the VPA-mortality is correct, but performs "badly" when it is set too low. This is quite as expected from the probing scenarios, as described in an earlier section.

The F_{low}-strategy performs badly in all the simulated scenarios, giving smaller and smaller catches while the cod stock grows. The growing cod stock is inevitable for the F_{low}-strategy, but with a growing cod stock the quotas should be increasing (or at least not be decreasing) even when the F_{low}-strategy is used to manage the stock. The explanation for this was found through simulating probing scenarios. When the spawning stock (SSB)— recruitment(R) relation is studied over a period with only large stock values, the resulting F-values (F_{low}, F_{med}, F_{high}) will be low, and the stock increases steadily. Putting some boundaries on the SSB—R relation, are sufficient to prevent this bad behaviour for the F_{med} and F_{high} strategies, but with the conservative F_{low}-strategy this is not enough when other experimental factors are introduced. When all experimental factors (except the management strategies) are on reference level, the F_{low}-strategy performs as expected.

The F_{high}-strategy gives smaller total catches than the F_{med}-strategy, but in the present parameterization of the simulation model also this strategy results in an increasing cod stock in several scenarios. This might be an indication that recruitment or growth (or both) is set too high. The performance of the F_{low}-strategy in this uncertainty experiment is an excellent example that one must keep the "common sense" in management considerations, and not leave all decisions to automated rules. The steadily decreasing cod quotas would have been prevented if knowledge concerning stock size had been taken into account. An increasing cod stock will be observed both by the fishermen, and during the research surveys, and so proper knowledge will be available from several sources.

Before turning to the capelin and herring catches, we will take a look at the reported cod catches. The most conspicuous difference to the total catches is the importance of the factor O (the overfishing). This is as expected, because it is only when factor O is on alternative level that the two responses differ. The fitted model

$$\mathrm{Re}\,ported\ \ catch\tilde{\ }SCOD*(M+O+RCOD+U+I)$$

suits the data well, having a multiple R-squared of 0.99. When all factors are on reference level, the predicted value in this model is not significantly different from the model giving the total catch. Also the main effects from alternative management strategies, and the interaction effects included in both models are quite similar. The capelin and herring strategies are not significiant for the reported catches, neither.

Over-fishing results in lower reported catches (quotas), but total catch remains approximately the same (factor O is not significant in the total catch model). Assuming that the fishermen split into two groups, those over-fishing the quota and those not over-fishing the quota, the result is that the law-abiding group gets lower catches and the other group gets higher catches than if all the fishermen kept to their quotas. The total quantity caught will, however, be approximately the same in both scenarios. In the present simulation model, we have not implemented economic considerations or risk calculations connected to the (possible) over-fishing or discards. Inclusion of such factors might affect the above conclusion, however. Further, the simulations tell that the management strategies are affected by the nominal VPA-mortality. This is not surprising, both due to the results from the probing scenarios and also because this quantity plays a main role in the VPA-procedure used for computing the quotas.

On the basis of the two estimated models, one can compute predicted values for any combination of the experimental factors. It is a total of 2^{10} combinations of factors, when excluding the management strategies. Figure 15 is a distribution plot of the predictions over the 1024 possible factorial combinations, and Table 20 gives some summary statistics on those distributions. From the prediction plot in Figure15, we observe that the strategies F_{med}, F_{high} and F=0.26 perform quite similarly. The median value over all predictions is approximately the same, and the variance is much higher for the reported catches than for the total catches. The constant F-strategy has, however, higher variances. Further we observe that the F_{max}-strategy performs approximately similar to the three strategies mentioned above on the reported catch, but does have the largest variance on the total catch, and the median value also lies outside the range for F_{med}- and F_{high}-predictions. The F_{low}-strategy performs worst, without any doubt. From the distribution plots, the two strategies F_{med} and F_{high} seem to be about equal, and when only the total and reported cod catches are defined as

Table 20. Some summary statistics on the prediction distributions for mean yearly cod catches, over the reality spanned by the biological and human experimental factors. Unit is mill.tonnes.

		Cod management strategy				
RESPONSE		F_{med}	F_{high}	F_{low}	F_{max}	F=0.26
Rep.	Median	0.327	0.312	0.148	0.295	0.307
Cod	5% quantile	0.202	0.216	0.094	0.143	0.160
Catch	95% quantile	0.451	0.408	0.203	0.448	0.454
	Range (90%)	0.248	0.192	0.109	0.305	0.294
Total	Median	0.422	0.398	0.183	0.346	0.398
Cod	5% quantile	0.390	0.368	0.098	0.222	0.341
Catch	95% quantile	0.454	0.427	0.269	0.470	0.456
	Range (90%)	0.064	0.059	0.171	0.248	0.115

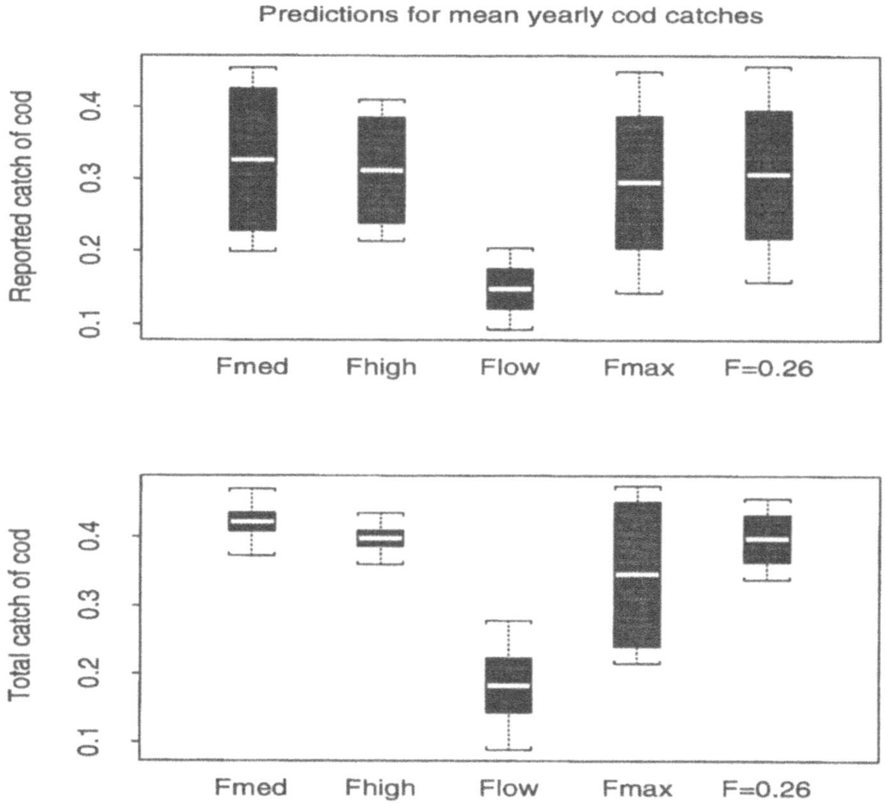

Figure 15. Plots illustrating the prediction distribution over the biological and human experimental factors. Unit on vertical axis is mill. tonnes.

performance measures, we would recommend one of those two. The median of the F_{med} prediction distribution is higher than that of the F_{high} distribution, and thus F_{med} might be the preferred strategy.

<u>Capelin</u>

For capelin all catches are reported, since there are no discards and no over-fishing included in the experiment. Figure 16 shows a marginal plot for mean yearly catch. The most important factors seem to be SCOD, M and RHER. However, the herring recruitment factor (RHER) does not give the expected effect. Lower herring recruitment (RHER on alternative level) was expected to result in a higher capelin stock and thus also a higher catch of capelin. The marginal plot indicates, however, that reduced herring recruitment results in a lower catch of capelin, and this was quite unexpected. Some additional simulations were run with all factors on reference level (10) and with only recruitment of herring on alternative level (10). The same effect appeared, with

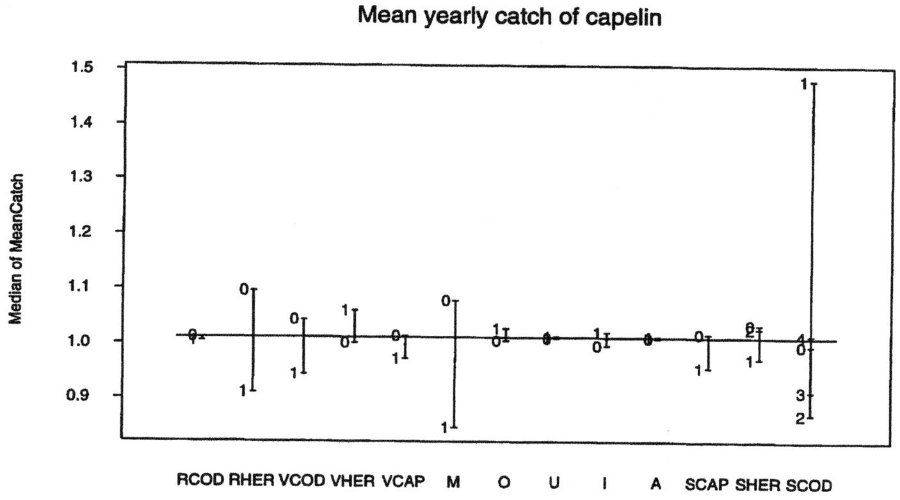

Figure 16. Marginal plot with catch of capelin as response variable. The vertical axis indicates median value when the experimental factors are on given level.

lower herring recruitment the mean catch of capelin was reduced. Taking a closer look on the simulations, we found that the reduced herring recruitment did result in a higher capelin recruitment, as expected. However, when the herring recruitment was on reference level ("high"), the capelin did get a higher growth, and this probably more than compensated the lower number. One additional explanation to the low capelin catches combined with the low herring recruitment, could be increased predation from the cod stock, since the availability of herring as food for cod is reduced when the herring recruitment is reduced. The increased predation from cod might be of greater consequences for the capelin stock than the predation from the herring.

A model fitting based on the same principles as for the cod catches, results in a simple linear model. To account for increased variance at higher catch levels, the catches were logarithmically transformed. The following model is preferred (multiple R-squared 0.777):

$$\log(Mean\ \ Catch)\tilde{}\ SCAP{*}M + SCOD{*}(RHER + U) \tag{8}$$

The interaction (U,SCOD) is significant only at 10% level. However, since discard of small cod should affect the cod-predation on capelin, we choose to include the interaction.

Table 21 shows the coefficient estimates for the preferred model. In this table one must remember that the response variable in the model is the log-catches, and so the predicted value when all experimental factors are on reference level is computed from the intercept term as

$$\exp(\text{int}\,ercept) = \exp(0.048) = 1.05\ \ mill.tonnes.$$

If the capelin strategy (SCAP) and the VPA-mortality for cod (M) both are on alternative level, then the predicted value is

$$\exp(\text{int}\,ercept + SCAP + M + SCAP*M) = \exp(0.048 - 0.043 - 0.150)$$
$$= 0.729 \;\; mill.tonnes.$$

The sign of the coefficient tells whether or not the effect contributes to an increase (positive sign) or a decrease (negative sign) in mean catch relative to the reference value. In Table 21, M has the largest negative effect, and SCOD on level 1 (strategy F_{high} for cod) has the largest positive effect on the capelin catches. The interactions (SCAP,M) and (SCOD-1,RHER) are highly significant, while the interactions (SCOD-3,RHER) and (SCOD-1,U) are significant on 10% level. It is a little surprising that the management strategy for capelin (SCAP) does not have a significant main effect (p-value 0.2). We suspect that this is due to the enormous natural fluctuations

Table 21. Coefficient estimates, model (8). Unit is log(mill. tonnes).

RESPONSE: LOG MEAN YEARLY CATCH OF CAPELIN

Main effects

Factor	Level	Value	Std. Error	P-value
Intercept		0.048	0.040	0.230
SCAP	1	-0.043	0.034	0.204
M	1	-0.171	0.044	0.000
RHER	1	-0.122	0.051	0.018
SCOD	1	0.433	0.054	0.000
	2	-0.046	0.060	0.452
	3	0.143	0.066	0.033
	4	0.028	0.048	0.563
U	1	0.103	0.064	0.110

Two-factor interactions

First factor	Second factor (level)	Value	Std. Error	P-value
SCAP	M (1)	-0.150	0.056	0.008
RHER	SCOD(1)	0.124	0.099	0.213
	SCOD(2)	0.199	0.076	0.010
	SCOD(3)	-0.162	0.095	0.093
	SCOD(4)	0.067	0.080	0.400
U	SCOD(1)	-0.149	0.085	0.082
	SCOD(2)	-0.029	0.085	0.730
	SCOD(3)	0.051	0.097	0.597
	SCOD(4)	-0.138	0.094	0.148

in the capelin stock. It is, nevertheless, interesting that reducing the spawning target from 500 000 tonnes to 100 000 tonnes does not, in our scenarios, have any significant long-term effect on the catches. SCAP is kept in the model though, because it interacts with M, and also because it is the factor of primary interest.

The cod-related factors SCOD (management strategy) and M (VPA-mortality) seem to have the largest effects on the mean capelin catch. The main effects of the SCOD-factor reflects the influence of cod fishery on the capelin catches. As one could expect, increasing the cod quotas (by using F_{high}-strategy instead of F_{med} for example), results in a higher catch of capelin as well. This is of course due to the reduced predation from cod, which is taken into account when the capelin quota is computed. Further, the nominal VPA-mortality for cod has considerable effect. Assuming this value to be 0.1 instead of the true value 0.2 results in lower capelin catches. Comparing with the effect of the M-factor on the total catches of cod (Table 19), we observe that this coincides with a lower predicted catch of cod, and so the reduced capelin quotas can be explained by increased predation.

We can observe that even if the capelin strategy does not show any main effect, it interacts significantly with the VPA-mortality factor M. When both SCAP and M are on alternative level (that means reduced spawning target for capelin, and too low nominal VPA-mortality for cod), the capelin catches are further reduced. From this, one might suspect that the increased predation from cod together with the reduced spawning target for capelin, will result in a reduced capelin recruitment and then also in reduced capelin catches.

From the table of coefficient estimates (Table 21) we observe that the main effect from reduced herring recruitment in fact is reduced capelin catches. As discussed above this is unexpected, but it might be explained by reduced individual growth and/or increased predation from cod. Looking at the interaction effects (RHER,SCOD), we observe that only when applying the F_{max}-strategy in the cod management the reduction in capelin catch is considerable. We then might suspect that an increased predation from cod is due to an interaction effect between the reduced herring recruitment and the cod strategy. The discard factor in cod fishery (U) results in slightly higher capelin catches, but when applying the F_{high} or constant F strategy in cod management this effect is nullified. The general increase can be explained by a reduction of the cod stock as a consequence of the discards. It is not clear why the discards in fact have the opposite effect when applying the F_{high}-strategy or the constant F strategy in the cod management.

Using the fitted model we can predict the mean yearly capelin catches over the response period for all combinations of the experimental factors. The capelin catches are affected by the cod management, and so there are 10 combined management strategies to judge. Figure 17 illustrates the prediction distribution over all 1024 combinations of the experimental factors, and for all the 10 combined management strategies. From this figure it is obvious that using F_{high} as cod strategy is favourable for the capelin fishery. It is also clear that it is not favourable to reduce the spawning target in the capelin strategy from 500 000 tonnes to 100 000 tonnes.

We can also observe that the F_{max}-strategy in cod management results in the highest variance in the prediction distribution for the capelin catches. Table 22 lists summary

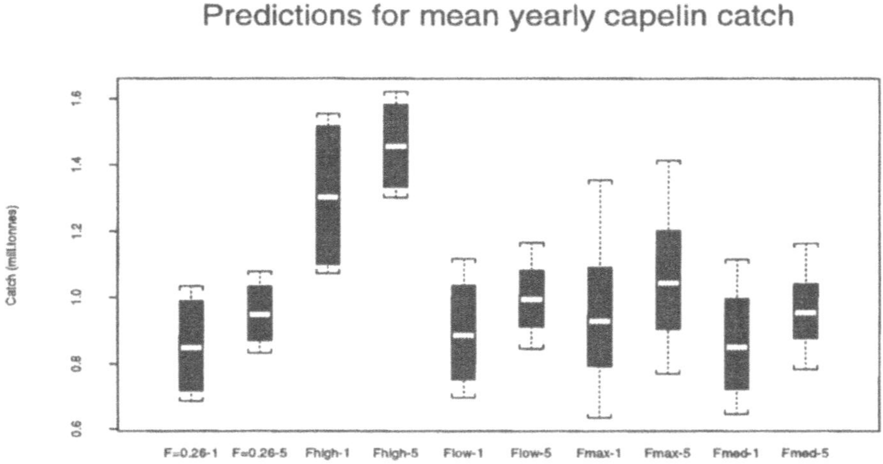

Figure 17. Plot illustrating the prediction distribution over the biological and human experimental factors. Unit on vertical axis is mill. tonnes. Horizontal axis labels the management categories SCOD-SCAP. Fhigh-1 means F_{high} strategy for cod and spawning target 100 000 tonnes for capelin, Fmax-5 means F_{max} strategy for cod and spawning target 500 000 tonnes for capelin etc.

statistics from the prediction distributions. From this it is obvious that the best management strategy when capelin catch is the performance measure, is a combination of F_{high} for cod and a spawning target of 500 000 tonnes for capelin. Since F_{high} did perform almost as well as F_{med} for the cod catch performance variable, this combination is attractive also when cod catches and capelin catches simultaneously are used as a performance measure.

Table 22. Some summary statistics on the prediction distributions for mean yearly capelin catch, over the reality spanned by the biological and human experimental factors. Unit is mill. tonnes.

Capelin strategy		Cod management strategy				
		F_{med}	F_{high}	F_{low}	F_{max}	F=0.26
Spwn.tar.	Median	0.955	1.455	0.993	1.042	0.948
500 000	5% quantile	0.783	1.301	0.845	0.768	0.832
tonnes	95% quantile	1.163	1.621	1.166	1.413	1.079
	Range (90%)	0.381	0.320	0.321	0.645	0.248
Spwn.tar.	Median	0.849	1.303	0.885	0.928	0.848
100 000	5% quantile	0.645	1.073	0.697	0.633	0.686
tonnes	95% quantile	1.115	1.553	1.117	1.354	1.034
	Range (90%)	0.469	0.480	0.421	0.720	0.348

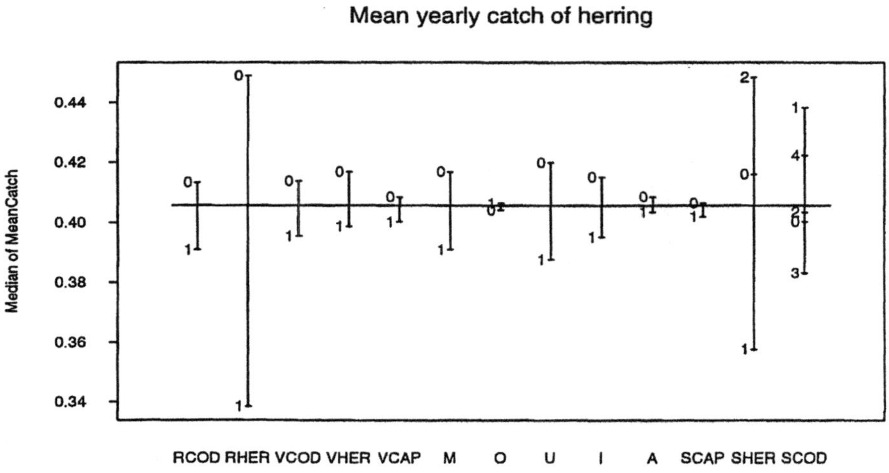

Figure 18. Marginal plot with catch of herring as response variable. The vertical axis indicates median value when the experimental factors are on given level. The unit is mill. tonnes.

Table 23. Coefficient estimates, model (9). Unit is log(mill.tonnes).

RESPONSE: LOG MEAN YEARLY CATCH OF HERRING

Main effects Factor	Level	Value	Std. Error	P-value
Intercept		-0.804	0.020	0.000
SHER	1	-0.089	0.024	0.000
	2	0.063	0.024	0.009
RHER	1	-0.254	0.020	0.000
VHER	1	0.032	0.020	0.105
U	1	-0.030	0.020	0.132

Table 24. Some summary statistics on the prediction distributions for the mean yearly herring catches, over the reality spanned by the biological and human experimental factors. The unit is mill. tonnes.

		Herring management strategy		
		$F_{0.30}$	$F=0.12$	$F=0.20$
Mean	Median	0.396	0.363	0.422
Yearly	5% perc.	0.337	0.308	0.359
Catch	95% perc.	0.462	0.423	0.492
Herring	Range(90%)	0.126	0.115	0.134

Herring

Only young herring is periodically present in the Barents Sea. This part of the stock is not subjected to fishery, and so we do not expect the herring fishery to be strongly dependent on cod- and capelin related factors. Figure 18 illustrates this. The main factor is obviously the herring recruitment, with management strategies for herring and cod coming next. As with the capelin catches, it is clear that the log-catch should be used as response in the model fitting.

The preferred model becomes:

$$\log(Mean\ \ Catch)\tilde{}\ SHER + RHER + VHER + U\,, \qquad (9)$$

it has multiple R-squared 0.676, and coefficient estimates as given in Table 23. The factors VHER (variation in herring recruitment) and U (discards in cod fishery) are not significant on the common 5% level. They are, however, included since they are factors of primary interest.

Since the preferred model does not include any interaction effects, we can read directly from the coefficient estimates that F=0.20 performs best and F=0.12 performs worst, while $F_{0.3}$ lies in-between. Table 24 displays summary statistics in the prediction distributions, similar to the cod- and capelin tables.

Mature biomass

It does not seem reasonable to fit linear models to a response variable counting the number of critical years (that is the number of years with mature biomass lower than the defined critical limit). An alternative would have been to fit binomial models. However, we do not feel comfortable with this neither, and so we will just present the simulation results and add some comments.

Cod

Figure 19 shows that the mature cod stock only gets lower than the minimum level 0.2 mill.tonnes when F_{high} is used as cod strategy. A closer examination of those scenarios leading to years with mature cod biomass below 0.2 mill.tonnes, reveals that the only thing in common is the cod strategy. In fact, of 27 simulations with F_{high} as cod strategy a total of 20 (that is approximately 75%) did get at least one year with mature biomass lower that the critical limit. In addition, all the 9 scenarios with F_{high} as cod strategy were represented among those 20 simulations. We thus conclude that using F_{high} as cod strategy, results in a risk of periodically depleting the cod stock. This does not come as a surprise, because of the definition of the F_{high} strategy. What perhaps is a little surprising, is that neither of the remaining cod strategies seem to entail any risk of depleting the cod stock. This might be artefacts of the simulation model and its parameterization.

Judged on catches of capelin and cod, the F_{high}-strategy came out best. Whether F_{high} still is the preferable strategy, when the risk of cod depletion is taken into account, depends on how these competing criteria are weighted.

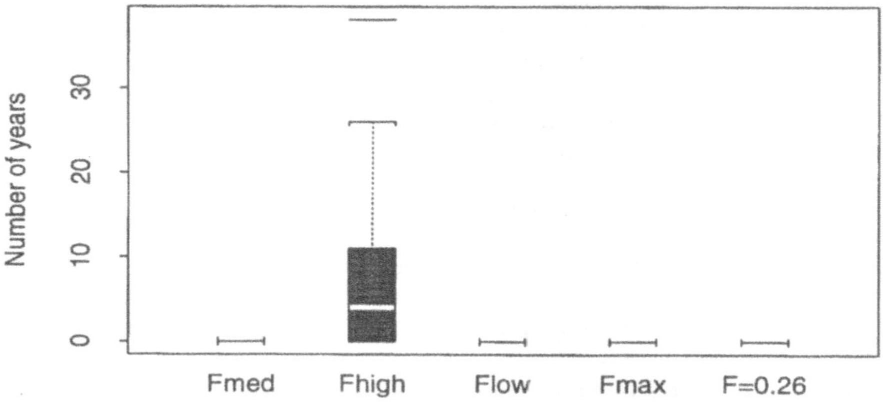

Figure 19. Marginal plot with number of years with mature cod stock lower than 0.2 mill. tonnes as response.

Herring

Using 1.0 mill.tonnes or 2.5 mill.tonnes as critical limits for mature herring biomass, does not give any information about whether or not the three herring strategies differ in risk of stock depletion. With the lower limit, only 1 out of the 129 simulations did result in a short period (3 years) with mature stock lower than this limit. With the larger limit, all simulations had periods with low mature stock, with the mean number of "critical" years during the response period being 43, 32 and 55 for the three herring strategies $F_{0.3}$, F=0.12 and F=0.20, respectively. Taking the intermediate critical value, 1.75 mill.tonnes, the F=0.20-strategy shows the worst results. 43 out of 45 simulations do have periods with mature biomass lower than the critical limit, with a mean of 23 "critical" years during the response period. The two remaining strategies perform approximately identical, with $F_{0.3}$ having a slightly higher fraction of simulations with critical periods. From these considerations, the $F_{0.3}$-strategy now seems to be the preferable herring strategy.

Discussion

From the cod and capelin catches, it is obvious that the combination of F_{high} for cod and a spawning target of 500 000 tonnes for capelin seems to be the best management strategy for the combined cod and capelin fishery. Using F_{low} as cod strategy constantly over a long period does not make sense, because the VPA-machinery soon starts to behave very badly. The F_{max} strategy results in very variable predicted cod catches, over the reality spanned by the factors in this uncertainty experiment. The strategies F_{med} and F_{high} performs equally, when only considering the cod catches over the response period. Introducing the capelin catches as an additional response, F_{high} is favourable to F_{med} as cod strategy because of the increase in capelin catches. This increase is due to the fact that F_{high} results in a higher fishing mortality than the F_{med}-strategy, and the cod stock is kept on a lower level. This leads to reduced predation from cod on capelin, and so more capelin is available for fishery. It does also seem clear that the spawning target in the capelin strategy should be 500 000 tonnes rather than 100 000 tonnes, even if the last one obviously results in an immediate gain for the fishermen. Over the total response period, however, it is no doubt in that the lowest spawning target also results in the lowest total catches. A possible explanation might be that the 100 000 tonnes target is not enough to give sufficiently recruitment in our simulation model, so there will be more years with no (or almost no) capelin fishery.

The herring catches are not affected neither by the cod strategy nor by the capelin strategy, and using the constant strategy F=0.20 gives the highest herring catches. The combination of F_{high} for cod, F=0.20 for herring and a spawning target of 500 000 tonnes for capelin then seems to be the best total strategy, when the combined fishery acts as performance measure.

Introducing some biomass considerations as additional performance measures, result in some modifications. The herring strategy F=0.20 weakens its candidature, because of the lowest score on mature biomass. The strategy $F_{0.3}$ does, however, strengthen its candidature to "best" herring procedure. The herring strategy $F_{0.3}$ also seems to contribute to stabilizing the cod catches, independent of cod strategy, compared to the F=0.20-strategy. The cod strategy F_{high}, however, weakens its candidature, because of the relatively large risk of damaging the cod stock. The capelin procedure is not judged during the stock considerations, since the mature capelin stock is not included as a response variable.

Figure 20 illustrates the results from one single simulation run of the scenario numbered 25 in the experimental design (Table 18). The large fluctuations in the mature herring stock are mostly due to the periodicity of extreme recruitment, and even if the mature biomass is on a minimum level at the end of the period, one will expect the mature stock to grow rapidly as soon as a year of extremely high recruitment has occurred. The fluctuations in the mature cod stock seem to be of smaller magnitude, but are, in fact, relatively larger than the herring fluctuations, as is seen from the log-scale plot in Figure 20. However, even if there are periods with low mature cod stock, the state of the stock seem to be quite stable on values just above the critical 0.2 mill.tonnes. This kind of performance is quite typical from the simulations with F_{high} as cod strategy. There are periods (of variable length) when the mature biomass is lower than the critical value of 0.2 mill.tonnes, but the values are fluctuating around the critical limit, and there are no signs (in these simulations) that the cod stock might suffer extinction if the F_{high}-strategy is used consequently.

Based on these reflections, it seems natural to draw the following conclusion based on the total catches and the mature biomass: The mixture of strategies F_{high} for cod, $F_{0.3}$ for herring and a spawning target of 500 000 tonnes for capelin is marginally best. This includes, however, an acceptance of keeping the cod stock on a low level, with the risk that the stock might be severely depleted if the recruitment fails for a long

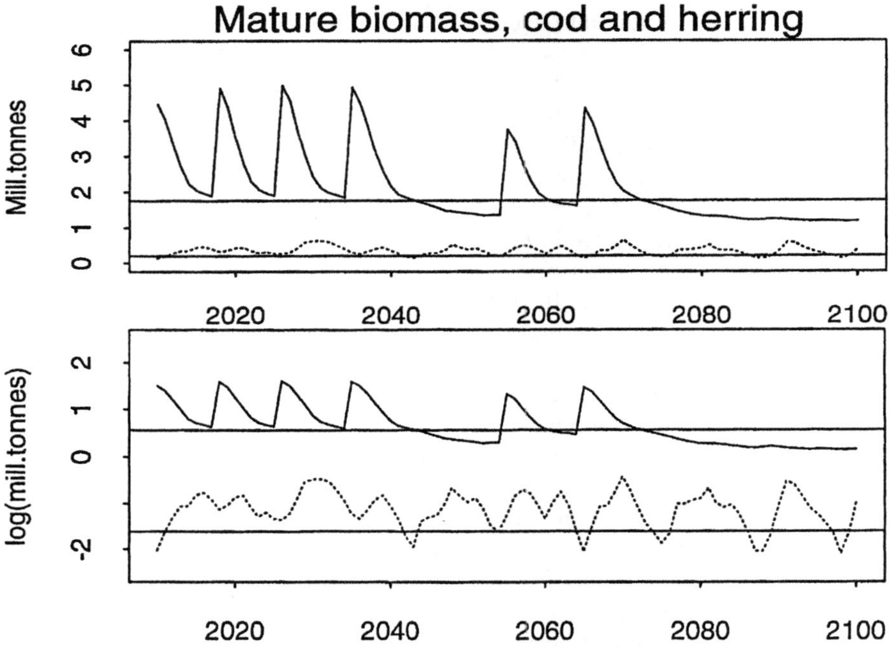

Figure 20. Results from one single simulation of the scenario no. 25 (see Table 18). The upper plot is on original scale, while the lower is on log-scale. In both, the upper solid curve shows the mature herring biomass, while the lower dotted curve shows the mature cod biomass. The horizontal lines are the critical values 1.75 mill. tonnes (herring) and 0.2 mill. tonnes (cod).

period. If one is not willing to accept this risk, changing to F_{med} will give approximately the same cod catches, but some 40% lower capelin catches.

APPENDIX

Predation and growth for cod

The predation model is developed for cod predating on capelin, herring and young cod. Modelling is done in three steps. First total stomach content is made dependent on relative abundance of the various stock components. Then the distribution of stomach content is made dependent on the same relative abundance vector. Finally relative and total stomach contents are converted to realized predation. A detailed description of data, analysis and resulting model is given in Hagen *et al.* (1995).

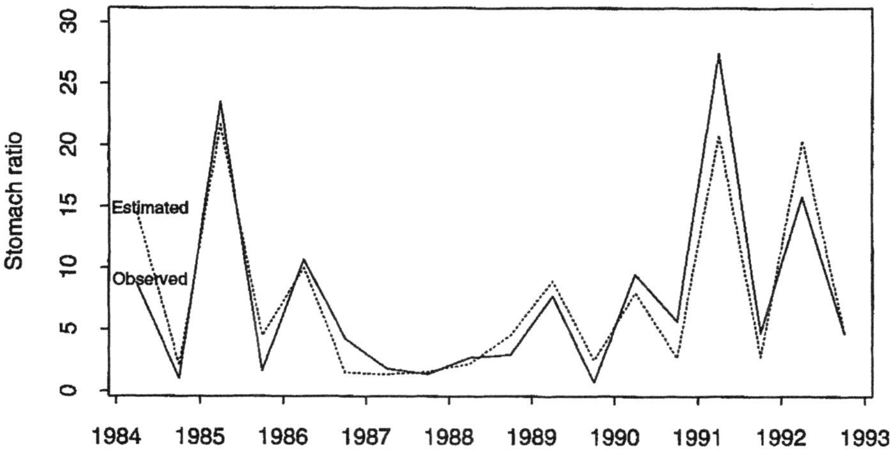

Figure 21. Stomach content of 3-9 year cod, observed and estimated values.

Data

The data are from the period 1984-92, and have been collected twice a year. Stomach data for cod and stock data for capelin, herring and cod have been used in the analysis. Since 1-2 year old cod is a potential prey for older cod, and 1 and 2 year old cod have different predation patterns, the cod stock is divided in three components, and so is the capelin stock. The total stock structure is

CAP_1	% immature capelin which is shorter than 10 cm
CAP_2	% immature capelin which is longer than 10 cm
CAP_3	% mature capelin
HER	% herring
COD_1	% 1 year old cod
COD_2	% 2 year old cod
COD_3	% 3 year and older cod

The percentages are computed according to the number of fish (and not biomass), and adding the 7 units in the above list will always yield the result 100.

Stomach fill

Stomach fill for 3-9 year old cod, R_3, is assumed dependent on relative stock abundance and season (H=0 for 1. halfyear, H=1 for 2. halfyear). The model is:

$$\begin{aligned} R_3 = A_0 + A_1 \cdot \log(CAP_1) + A_2 \cdot \log(CAP_2) + \\ A_3 \cdot (1-H) \cdot \log(CAP_3) + A_4 \cdot \log(HER) + \\ A_5 \cdot \log(COD_1) + A_6 \cdot (1-H) \cdot \log(COD_1) + \\ A_7 \cdot \log(COD_2) + A_8 \cdot \log(COD_3) \end{aligned}$$

Stomach fill is defined as a fraction: weight of total stomach content to the total weight of the cod. The parameters in this model are estimated by linear regression ($R^2 = 0.73$). Observed and predicted stomach fill (%) is plotted in Figure 21.

Stomach fill for 1-2 year old cod is assumed to follow the same model, and are also estimated by linear regression ($R^2 = 0.56$).

Stomach composition

Stomach content is decomposed into 6 groups:

$PCAP_1$	immature capelin shorter than 10 cm
$PCAP_2$	immature capelin longer than 10 cm
$PCAP_3$	mature capelin
PHER	herring
$PCOD_1$	1 year cod
$PCOD_2$	2 year cod

Similar to the data, the components are measured in relative units (% of total). Let P be any one of these prey components. Then P is modelled as:

$$\begin{aligned} P = {} & A_0 + A_1 \cdot \log(CAP_1) + A_2 \cdot \log(CAP_2) + \\ & A_3 \cdot (1 - H) \cdot \log(CAP_3) + A_4 \cdot \log(HER) + \\ & A_5 \cdot \log(COD_1) + A_6 \cdot (1 - H) \cdot \log(COD_1) + \\ & A_7 \cdot \log(COD_2) + A_8 \cdot \log(COD_3) \end{aligned}$$

The coefficients are estimated separately for each component, without ensuring that the components sum to the total. In the simulation model negative predicted percentages are replaced by 0, and the resulting sum is scaled to 100.

Predation mortality

Predicted stomach fill and stomach composition are converted to consumed total, according to dos Santos and Jobling (1992) and an unpublished work of dos Santos from 1994:

$$C_i = \frac{\ln 2 \cdot S_i \cdot e^{0.13 \cdot T} \cdot B^{0.26}}{\alpha_i \cdot S_o^{0.52}}$$

C_i	is the weight of the consumption of prey species i
T	is the temperature
B	is the weight of cod
S_i	is the relative stomach content of prey component i
S_0	is the stomach fill
α_i	is a prey specific constant which indicates the halfvalue time for prey species i

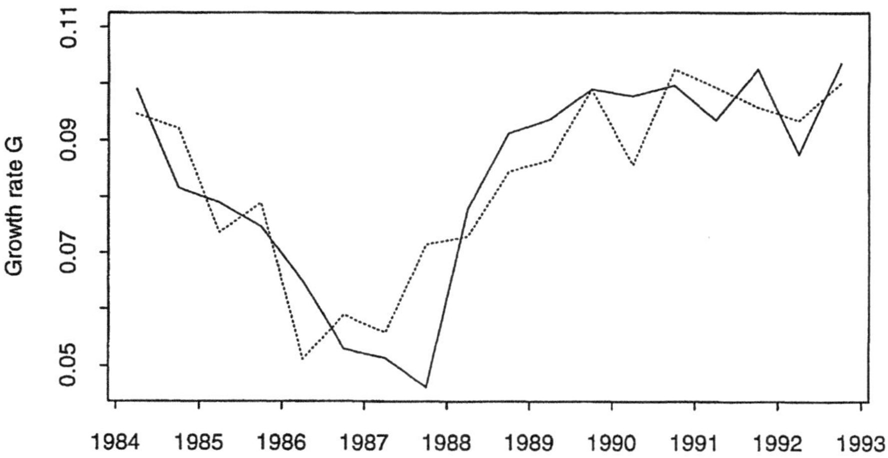

Figure 22. Observed (solid line) and estimated values (dotted curve) for the cod growth model.

The consumed totals are subtracted from the various stock components as predation mortality.

Individual growth

Individual length growth for cod is assumed to be a linear function of the relative abundance vector:

$$G = A_0 + A_1 \cdot COD_1 + A_2 \cdot COD_2 + A_3 \cdot COD_3 + A_4 \cdot CAP_1 + A_5 \cdot (CAP_2 + CAP_3) + A_6 \cdot HER$$

The growth rate G is defined by $G = (L_{t+T} - L_t) / (L_\infty - L_t)$.

L_t denotes the cod length at time t, T denotes the interval length, and L_∞ is the maximum length of the cod. In the computations L_∞ is set to 160 cm.

The model is fitted by linear regression separately for each of the first 6 age classes of cod. Observed and predicted aggregated linear growth is shown in Figure 22.

REFERENCES

von Bertalanffy, L. 1938. A quantitative theory of organic growth. Hum.Biol. (10)2: 181-213.

Bogstad, B., H. Gjøsæther & S. Tjelmeland 1992. CapTool - a versatile aid in Barents Sea capelin catch quota options calculation. Working document to the 1992 meeting of the ICES Atlanto-Scandian Herring and Capelin Working Group.

Bogstad, B. & S. Tjelmeland 1993. A method for estimating predation mortalities on capelin using a cod-capelin model for the Barents Sea. Unpublished manuscript.

Box, G. E.; W. G. Hunter & S. Hunter 1978. Statistics for Experimenters. J. Wiley& Sons, New York.

Chambers, J. M. & T. J. Hastic (eds.) 1992. Statistical models. In S. Wadsworth & Brooks / Advanced Books and Software, Pacific Grove, California.

Cooke, J.G. 1995. The International Whaling Commission's Revised Management Procedure as an example of a new approach to fishery management. Pp. 647-657 in: Blix, A. S.; L. Walløe & Ø. Ulltang (eds.). Whales, seals, fish and man. Elsevier Science B.V.

dos Santos, J. & M. Jobling 1992. A model to describe gastric evacuation in cod (*Gadhus morhua* L.) fed natural prey. ICES J. mar. Sci. 49: 145-154.

Hagen, G.; E. Hatlebakk & T. Schweder 1994. SCENARIO BARENTS SEA. A tool for evaluating Fisheries Management regimes. NR-note STAT/07/1994.

Hagen, G.; E. Hatlebakk & T. Schweder 1994.SCENARIO BARENTSHAV: Simuleringsresultater (1) (in Norwegian). NR-note STAT/08/1994.

Hagen, G; E. Hatlebakk & T. Schweder 1995. SCENARIO BARENTSHAV, Predasjonsmodell (in norwegian). NR-note STAT/02/1995.

Hilborn, R. 1992. Current and future trends in fisheries stock assessment and management. S. Afr. J. mar. Sci. 12: 975-988.

Hylen, A., O. Nakken & K. Sunnanå 1986. The use of acoustic and bottom trawl surveys in the assessment of north-east arctic cod and haddoc stocks. Alton, M. (ed). A workshop on comparative biology, assessment and management of gadids from the North Pacific and Atlantic oceans. Northwest and Alaska Fisheries Center, Seattle. p. 473-497.

ICES. 1993. Report of the Arctic Fisheries Working Group Copenhagen 1993.

IWC. 1993. Report of the Scientific Committee, Rep. Int. Whal. Commn 43, 57-64

Jakobsen, T. 1992. Biological reference points for North-East Atlantic cod and haddock. ICES J. mar. Sci. 49: 155-166.

Laurec, A. & J. G. Shepherd 1983. On the analysis of catch an effort data. J. Cons. int. Explor. Mer. 41: 81-84.

Pope, J. G. & J. G. Shepherd 1982. A simple method for the consistent interpretation of catch-at-age data. J. Cons. int. Explor. Mer. 40: 176-184.

Pope, J. G. & J. G. Shepherd. 1985. A comparison of the performance of various methods for tuning VPAs using effort data. J. Cons. int. Explor. Mer. 42: 129-151.

The Use of Multi-Species Models in Fishery Management

By Per Sandberg.
Fiskeriøkonomisk avdeling, Fiskeridirektoratet, Boks 185, 5002 Bergen, Norway.

In this chapter, we will discuss how the fishery management in Norway may use multispecies models when regulating fisheries in the Barents and Norwegian sea. We will start by giving a short description of the main fisheries in these areas, and discuss which important management questions the models should shed light on. Thereafter we will show how the fishery management solve these questions, and at which level in the process knowledge from the models may be implemented. In the end, we will point out how multispecies models may be of use to the managers in the future. This chapter was written in spring 1995, and consequently refers to the management conditions at that time.

THE FISHERIES IN THE BARENTS AND NORWEGIAN SEA

In addition to the North sea, the Barents and the Norwegian sea are the most important areas for Norwegian fishermen. Here fishermen catch cod, haddock, saithe, shrimp, herring, capelin, minke whales, seals and various other species. The multispecies models have so far focused upon the core stocks cod, herring and capelin, and we will give a short presentation of each of these fisheries.

Northeast Arctic cod

The cod stock has a dispersal from the south-west coast of Norway in the south to the Spitzbergen and Novaja Zemlja islands in the north and east, and is caught in most of this area (see Figure 1.).

The stock has a zonal attachment to both the Russian and the Norwegian economic zones, as well as to the Fishery protection zone around Svalbard, and is thus transboundary. Sometimes a minor part of the stock is also found in the high seas between the above mentioned areas. The total allowable catch (TAC) for cod has, since the establishment of the economic zones, been fixed by The Joint Norwegian Russian Fishery Commission. After allocations to third parties (EU, Faroe Islands, Greenland and Poland) the residual is shared 50/50 between Norway and Russia. Figure 2. shows the Norwegian quota for the Northeast Arctic cod stock in the period

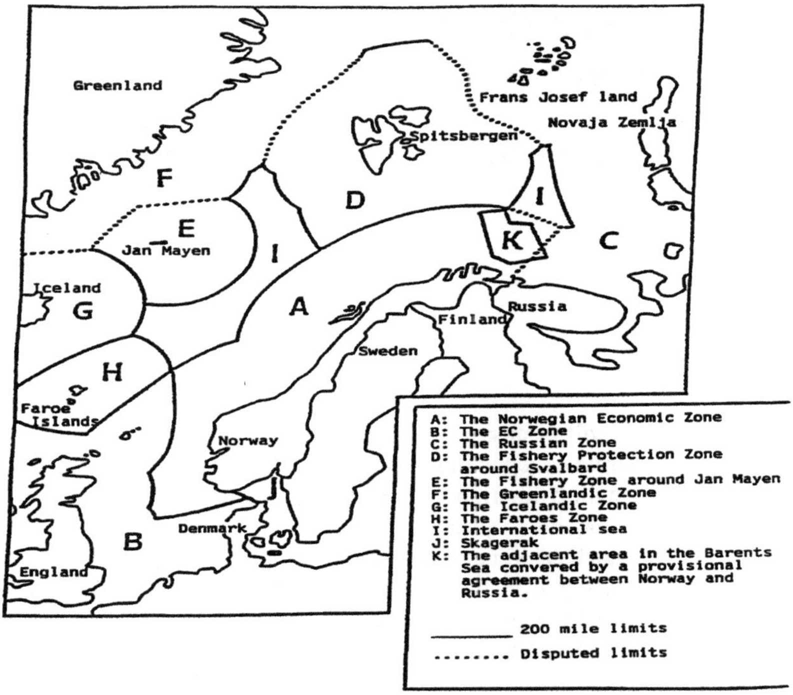

Figure 1. Map of Barents and Norwegian sea (Engesæter, 1991).

1980 - 1995.

The Norwegian quota was reduced from 343 000 tonnes in 1987 to 113 000 tonnes in 1990, a reduction of approximately 67% (The Joint Fishery Commission). This reduction constituted a hard burden on Norwegian fishermen and were the main reason for introducing individual vessel quotas in the coastal fleet in 1990.

During the last years, approximately 8 000 vessels fishing with passive gears (nets, long line, danish seine) and 120 trawlers have participated in the Norwegian cod fishery. The trawlers are regulated with individual vessel quotas. During the period 1990 - 1994 the most efficient of the vessels fishing with passive gears was regulated with individual vessel quotas, but from 1995 all vessels in this group are participating in a more competitive fishery where each vessel is given a non guaranteed quota (i.e. the fishermen risk a stop in the fishery before they have caught their individual quota).

Cod is the most valuable specie for Norwegian fishermen, and the value of the catch reached 2.7 billion NOK in 1994. This amounts to 37% of the first hand value of total Norwegian fisheries. Through the large number of participants it plays a key role for the economy of many coastal communities.

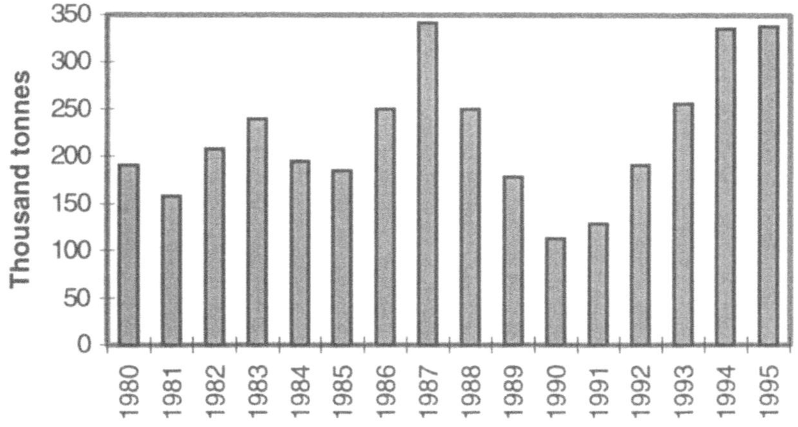

Figure 2. Norwegian quota for the northeast Arctic cod stock in the period 1980 -1995 (coastal cod included).

Norwegian spring spawning herring

The stock of Norwegian spring spawning herring has a dispersal from the south-west coast of Norway in the south to the Barents sea in the north. As cod, it is transboundary and has a zonal attachment mainly to the Norwegian but also to the Russian waters. During the 1950s and the 1960s a large fishery led to a collapse of the stock, and since 1970, the fishable part of the stock has been concentrated in the Norwegian Economic Zone (NEZ). Strict regulations in the 1970s and 1980s have been the managers contribution to securing the spawning stock. Good recruitment the latest years has made the stock recover, and the International Council for the Exploration of the Seas (ICES) has, depending on the level of TAC in 1995, assessed the stock to between 3.5 and 4.0 million tonnes by the end of 1995 (Anon 1995). Figure 3. shows the Norwegian quota for Norwegian spring spawning herring in the period 1980 - 1995.

The recovery of the stock is expected to influence its dispersal in the Norwegian Sea. For the first time in 25 years part of the stock was observed and caught outside NEZ in 1994. This fishery on the high seas has continued in 1995, and must be regulated in order to avoid a overfishing of the stock. If the stock follows the migration pattern of the 1950s, it will be available in both the Icelandic and the Faroe Islands economic zones as well as in international waters. This will give rise to new regulatory arrangements. However, in the following we will discuss management as it has been until 1994, that is, an arrangement where Norway have fixed the TAC and have allocated parts of this TAC to Russia.

During the last years, the TAC for Norwegian spring spawning herring has been shared with approximately 85% to Norway and the rest to Russia. Approximately 800

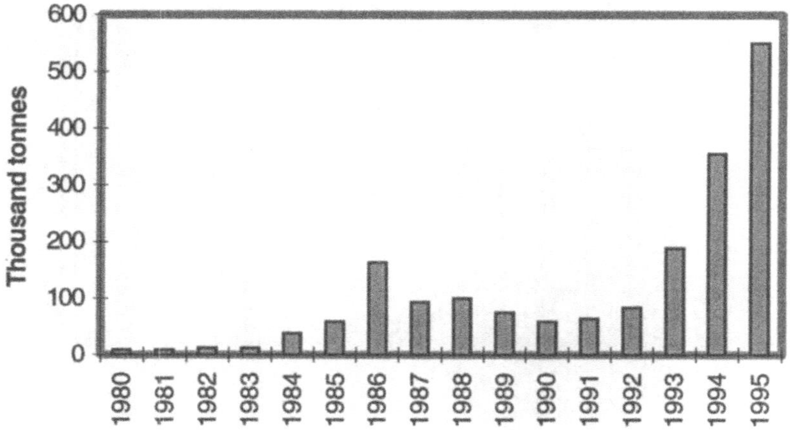

Figure 3. Norwegian quota for the Norwegian spring spawning herring in the period 1980 - 1995.

vessels participate in the Norwegian fishery. 100 of these are large purse seiners, 70 are trawlers and the rest are smaller seiners operating in coastal areas. The large purse seiners are regulated through individual vessel quotas, whereas the trawlers and the coastal seiners are participating in a more competitive fishery like the one described for cod.

Barents sea capelin

The name of this stock indicates its dispersal. Alike the two other stocks, it is transboundary and has a distribution in the Norwegian and the Russian economic zones, and in the Fishery protection zone around Svalbard. The TAC for capelin is shared with 60% to Norway and 40% to Russia. The specie has a short life-cycle, and the size of the biomass shows great variability during short periods of time. This is reflected in Figure 4. which shows the Norwegian quota for the Barents sea capelin in the period 1980 - 1995.

The yield from the capelin stock shows even greater variations than the yield from the cod stock. The stock was nearly wiped out in 1986, resulting in a moratorium for the years 1987-1990. The fishery was reopened in 1991, but closed again in the autumn of 1993. At present (1995), there are no signs of recovery for this stock.

During the capelin fishery in the years 1991 - 1993, approximately 210 Norwegian vessels participated in the fishery. This fleet consists of roughly 100 large purse seiners, 40 trawlers and 70 coastal seiners. The purse seiners are regulated through individual vessel quotas, whereas the remaining vessels are participating in a more competitive fishery. Nearly the entire Norwegian catch of capelin is delivered for

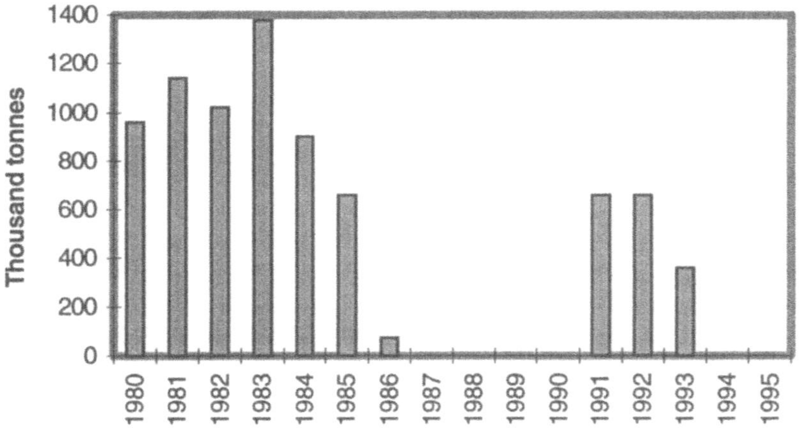

Figure 4. Norwegian quota for the Barents sea capelin stock in the period 1980 -1995.

reduction (production of fish meal and oil). There is also a small market for frozen capelin as well as roe for human consumption.

The multispecies dimension in the management of these resources is not a «mixed fishery» problem but a «species interaction problem». Although parts of the fleet operate in both cod and herring fisheries, or in both herring and capelin fisheries, the species are caught in separate operations. As the figures show, available quotas of the three species for Norwegian fishermen have shown considerable variations during the past 15 years. In particular, the reduction of the quota of cod in the period 1987 - 1990 and the moratorium of the capelin fishery in the same period had serious economic effect on Norwegian fishermen and processing industry.

These events were also the main arguments for initiating the «Multispecies research program» in 1990. It was common knowledge that cod preys on capelin and juvenile herring, but one was not able to quantify this interaction or to forecast them in time and area. The major aim of the program was to generate such knowledge. With this in mind, let us evaluate which management problems may benefit most from the models which the program has initiated.

RELEVANT PROBLEMS FOR MULTISPECIES MODELS

Managing the above mentioned fisheries in the Barents and Norwegian sea is characterised by solving numerous interrelated questions. Some of the most important of these are;

1. Choice of management strategy and the yearly level of TAC
2. Technical regulations (mesh size, minimum size, closed areas, etc.)
3. Division of the Norwegian quota on different parts of the fleet
4. Choice of how to regulate the different parts of the fleet (individual vessel quota, Olympic fishery, etc.)
5. Criteria for level of participation (how many vessels should be licensed to harvest a specific fish stock)

The stocks of cod and capelin are managed jointly by Norway and Russia through The Joint Norwegian - Russian Fishery Commission, whereas the herring stock until 1994 has been managed by Norway. This fishery commission will fix the TAC for cod and capelin and discuss the level of TAC for herring. It will further decide on various technical regulations and in many instances be the correct forum for question 1 and 2. The three last questions will be dealt with unilaterally by the Norwegian Fishery Administration.

With their focus on species interaction, the multispecies models should help managers to find better solutions to question 1: choice of management strategies and the yearly level of TAC for the three species. In addition, they will be of help when discussing question 2: technical regulations. Technical regulations which influence the age structure and size of the biomass for one specie will, when the species interact, also influence the others. If knowledge stemming from the multispecies models can reduce the risk of overexploiting the fish stocks and stabilise the yield from them, the management would have been brought a step forward.

Multispecies models will also be useful when solving the last three questions. We find, however that less advanced models may help to solve such questions, and that running a multispecies model each time these items are discussed may be unnecessary cumbersome.

Consequently, the fishery management has two important tools when conducting «multispecies management». That is, the TAC regulation and technical regulations. The technical regulations are regulations which traditionally have been stable over several years, whereas the TAC for cod, herring and capelin are fixed each year.

The level of TACs for the three species therefore seems to be the most immediate question which multispecies models should shed light on. Considering the distribution of the fish stocks, it is self-evident that the decision making process when fixing TACs must be international in its character. In order to get a picture of where results from the multispecies models can be implemented we shall give a brief description of how this decision making process works.

Fixing TAC - the decision-making process

The decision-making process when fixing the TACs follows this pattern each year:

1. As mentioned earlier, all three species have a large part of their distribution in the Norwegian economic zone and the Russian economic zone, although parts of the stocks also may be found outside the exclusive economic zones. However, considering the size of the distribution in Norwegian and Russian waters, these coastal states

are responsible for managing the fish stocks. The institutes of marine research in these two countries conduct the annual surveying of the stocks. During these surveys the size and composition of the stocks are measured.

2. Data from both surveys and the fisheries are annually assessed by the ICES' working groups, which presents the assessments in their various reports each year.

3. ICES' Advisory Committee on Fisheries Management (ACFM) reviews the reports presented by the working groups, and gives annual advice to fisheries management bodies of the coastal states on how much to fish each year. Experts from all ICES member countries are members in ACFM, and the advice given thus achieves international scientific acceptance. If the stock is above safe biological limits ACFM presents tables which show the consequences of different levels of fishing on biomass and spawning stock biomass with short term projections. If the stock is below safe biological limits, ACFM presents a more specific advice in order to rebuild the stock.

4. The advice or the options presented by ACFM will be analysed by the fisheries management bodies in Norway which try to find the level of fishing which seems most beneficial. The Norwegian position on this question is arrived at before Norway and Russia finally decide on the TACs, which thereafter is divided between Norway and Russia, and allocations are made for third nations.

The quota allocated to Norwegian fishermen is therefore fixed during the yearly fisheries negotiations with Russia. After this is done, the Norwegian quota is split between different vessel groups and individual vessels. The division of quotas between national fishermen does not have any direct consequences for the amount to be fished and is therefore of limited relevance in this chapter. Table 1 shows the decision-making process as described above.

So far, we have seen the following: The three species which the multispecies models have focused on are transboundary and the level of TACs builds to a great extent on the international accepted advice given annually by ACFM. When the stocks are outside safe biological limits, ACFM provides specific advice, and when the stocks are considered to be within safe biological limits, fishery managers will choose levels of TAC within the range of options given by ACFM.

When making such a choice, we need to know the objectives which should be fulfilled when managing the fish stocks.

Table 1. The decision-making process.

DATA OBTAINED FROM SURVEYS AND FISHERY
REPORT OF ICES WORKING GROUPS
ADVICE ON LEVEL OF FISHING (ACFM)
ANALYSIS OF ADVICE FROM ACFM
FISHERY NEGOTIATIONS WITH RUSSIA

OBJECTIVES WHEN FINDING MANAGEMENT STRATEGIES AND THE ANNUAL LEVEL OF TAC

Unlike the assumptions often made in bioeconomic textbooks, there does not exist a «sole owner» to the cod, herring and capelin stocks whose objective is to maximise profits from the resources. The stocks can to a certain degree be said to be owned by the Russian and Norwegian societies and, depending upon migration routes, to other nations or the international community. When searching for the objectives which the Norwegian society may have for the management of these resources, we need guidelines for the fishery policy drawn up by the Norwegian parliament. These are stated in several white papers, and the four most often quoted are the following (White Papers 1982 - 1991):

- The existing pattern of settlement shall be preserved.
- The marine resources shall give sustainable yields.
- The people shall have secure and good jobs.
- The value-added (profit and wage) in the fisheries sector shall be increased.

In general, it can be argued that these objectives cannot be fulfilled simultaneously. For example, if it is important to achieve a higher value-added in the fisheries sector, this might imply a regulation which reduce effort and which would not lead to a preservation of a given pattern of settlement. This non-compatibility of objectives makes it difficult to implement regulations which only fulfils one of the above mentioned goals.

When fish stocks are to be rebuilt to safe biological limits, a conflict between short term and long term management seems to appear. Numerous White Papers have emphasised that in such conflicts, long-run considerations should outweigh short run considerations. This implies that the objective stating that the marine resources shall give sustainable yields will be vital to be able to fulfil the other objectives.

Concerning the management of a fish stock with no interaction with other species, it can be argued that a highest possible *sustainable* TAC which does not put the stock outside safe biological limits should be the target. When two or more stocks interact, the harvest of each will influence the other, and when deciding upon the level of TAC for the different stocks, it is important to get a set of TACs which as a group fulfils the objectives mentioned above in the best way possible. That is: the combination of TACs for cod, herring and capelin should ensure the preservation of the existing pattern of settlement, sustainable yields, secure and good jobs for the people and increased value added in the fisheries sector.

Clearly, the fixing of a package of TACs for stocks which interacts, demands insights into the biology of the fish stocks as well as economic features of the harvesting sector, and the aim of the multispecies models is to develop such insights.

IMPLEMENTATION OF RESULTS FROM MULTISPECIES MODELS IN MANAGEMENT

This book has presented two biological multispecies models, an economic multispecies model and a scenario model. In addition, chapter 2 has outlined the most important oceanographic conditions in the Barents sea and their influence on the fish stocks.

In general, results from multispecies models may be implemented in fisheries management through the following two channels:

1. Existing advisory system (ICES - ACFM).
2. Through management strategies.

Let us look at these two channels.

Existing advisory system (ICES - ACFM)

Scientific results approved by the international scientific community will gradually be adopted by ICES working groups in their yearly assessments of the cod, herring and capelin stock. Although ICES still use single species models, estimates of recruitment, growth and natural mortality for these stocks may be improved when taking species interaction into account.

To managers, MULTSPEC is a scientific model which may produce such results. The strength of this model is that it is disaggregated and may be used when testing various alternative hypothesis on the natural and man-induced processes affecting the cod, herring and capelin stocks in the Norwegian/Barents sea. Results from work with the MULTSPEC model has already been used by ICES' working groups and implemented in ACFM's assessments:

• The management of capelin is based on a target escapement strategy. This implies that the quota on capelin should be adjusted to secure a spawning stock of about 500 000 tonnes. At present ACFM give advice on the level of TAC by producing a forecast of spawning stock biomass. The forecast is given ½ year in advance of the spawning, and the estimate of the spawning stock will depend heavily on the estimate of natural mortality of capelin during this period.

For many years the estimate of natural mortality on capelin was set irrespective of the size of the cod stock. As a result of the multispecies research, this estimate is now set after taking the food requirement of the cod stock into consideration.

• The estimates of weight at age for cod has changed from using a constant value to using a value dependent upon the size of the capelin stock.

We believe that management will benefit from the MULTSPEC model first and foremost by the scientific results which gradually will improve the stock assessments and the advice given by ICES/ACFM.

Work with management strategies

Fixing the yearly level of TAC's for the different stocks is a short term decision with long term consequences. The level of TAC will determine how much of the stock which will survive and be fishable in the future, and what level of TAC which may be set in the following years. In order to fix a TAC compatible with the objectives for the management of the stocks, these long term consequences should be known.

The fishery management in Norway has started to work with mapping such consequences and to evaluate which management strategy seems to be most compatible with the objectives for the management of the stocks (Nakken et.al.1994, Bogstad et. al. 1994). There seems to be two directions this work should follow:

1. As mentioned earlier, ACFM provides an assessment of the various stocks which forms a common understanding of the biology both by Norway, Russia and, dependent on migration pattern, other nations with interest in the fish stocks. When mapping the long-term consequences of different levels of TAC one should build on the same estimates of the stock as do the assessments presented by ACFM. This is important to achieve legitimacy of the forecasts.

But ACFM do not provide long term estimates of recruitment, growth and natural mortality for the stocks considered. In addition to the assessment provided by ACFM, fishery managers must therefore seek the best possible forecasts on how these parameters will evolve in the years to come. Having done this, the model AGGMULT and ECONMULT will be of particular use in simulation.

AGGMULT is an aggregated multispecies model. Contrary to MULTSPEC, it has not divided the different fish stocks on geographical areas nor has it as sophisticated length distribution. However, stock-recruitment relationships are included.

Due to its level of aggregation, AGGMULT may be used to simulate stock development for several years in response to different assumptions on recruitment, species interaction, TACs or fishing mortalities. Simulating with AGGMULT we may for example find which level of TAC or fishing mortality which, in the long run, will give the highest biological yield from the three stocks without pushing them outside safe biological limits. Results from such simulations may be used when formulating management strategies for the different fish stocks.

ECONMULT is a bioeconomic multispecies model and may be used to calculate the economic consequences for the Norwegian fishermen of different level of TAC on cod, capelin and herring. The model can be connected to any biological multispecies model, but has so far been connected to AGGMULT. ECONMULT may provide necessary forecasts of revenue, costs and division of these on different vessel groups. In particular it is an important tool for managers by describing the economic interdependencies between the different fish stocks.

2. However, the uncertainty with any long term estimate of the natural processes that determine the size of a fish stock should always be recognised. It is extremely difficult to present correct estimates on recruitment, growth or natural mortality years ahead. And it will be as difficult to present estimates of future prices of the fish and harvesting costs.

In addition to the work described above, one should therefore try to establish guidelines for a sound management when explicitly recognising uncertainty. This can be done by simulating a long period of time (30 - 100 years), and feed the simulation model with a range of probable estimates of recruitment, growth and natural mortality. Such a method will take care of the uncertainty inherent in all forecasts, and it will be possible to find the average exploitation pattern, spawning stock biomass, effort, etc. which in the long run will give the best performance according to the objective of the fishery policy.

The Scenario model is such a model. It consists of a simulation model, a management procedure and a set of performance measures. The model may improve management in several ways:

First, for a given management procedure, alternative specifications of the simulation model may reveal which factors are important when assessing the fish stock. This may indirectly lead to better assessment models.

Second, by establishing different management procedures, we can check which give the best performance. If, for example, a fishing pressure of F_{med} on cod, F=0.10 on herring and a target escapement of 500 000 tonnes of capelin gives the best solution according to some criteria, such a result should be used as guidelines when formulating the management strategies for the three stocks.

The different models mentioned above will help the fishery management in different ways. The complex MULTSPEC model for improving the existing assessment models, the AGGMULT and ECONMULT to simulate forecasts of biological and economic parameters to be used in management strategies, and finally the SCENARIO BARENTS SEA to establish guidelines for the management.

At present, our best suggestion for management strategies for the three core species is:

Cod	Minimum spawning stock of 500 000 tonnes
	Fishing mortality F= 0.20 - 0.40
Herring	Minimum spawning stock of 2 500 000 tonnes
	Stable TAC, in the level 600 000 - 800 000 tonnes
Capelin	Minimum spawning stock of 500 000 tonnes

These strategies represent a picture of the management of the stocks. They will be challenged in the years to come, and scientific results stemming from the multispecies models will undoubtedly be one of the driving forces in such a change.

Table 2. Relevance of multispecies models to fishery management.

Model	Relevance to fishery managers
MULTSPEC	Improved advice from ICES/ACFM
AGGMULT - ECONMULT	Management strategies - forecasts
SCENARIO MODEL	Management strategies - guidelines

FUTURE IMPLEMENTATION OF RESULTS FROM MULTISPECIES MODELS

The shortcomings of ACFM's use of single-species models when assessing stocks which interact is obvious: There are no guarantee that the presented options of TACs for each species will be internally consistent. And furthermore, the decision makers will not have the information at hand to judge the consequences different levels of TAC will have on the two other stocks.

It would therefore help if ACFM started giving advice built on multispecies models. Such advice should be given in form of tables where jointly compatible TACs for the species considered are shown. When ACFM accepts these models as "state of the art", they will form the biological basis for both Russia and Norway when they jointly are to determine the TACs.

So far the multispecies models have focused on cod, herring and capelin. In addition it would be useful to include sea mammals and shrimp. With the large biomasses of both harp seal and minke whales, it is important to know how much they eat of commercial valuable fish stocks. Flåten and Stollery has assessed the value of the fish eaten by the Northeast Arctic Minke whale, and found a net value of NOK 11 592 per whale per year (Flåten and Stollery, 1994). If this whale population is within the 95% probability range estimated by the International Whaling Committee's scientific body (61 000 - 117 000 animals), the economic effect of predation per year will be in the order 0.7 - 1.4 billion NOK.

The inclusion of the shrimp stock in the multispecies models is also important because cod, in particular, preys on shrimp, and the value of the Norwegian shrimp fishery is quite substantial (approximately 0.6 billion NOK per year).

CONCLUSIONS

In this chapter one of our aims has been to show the international character of the management the cod, herring and capelin fisheries in the Barents and Norwegian sea. ACFMs central position as an international neutral biological adviser is necessary in this process. When the stocks are within safe biological limits, fisheries management authorities in Norway will give advice on the level of TAC and such advice must be compatible with the assessments made by ACFM.

The models and in particular results from the models will hopefully lead to a higher quality of the management of the fish stocks in the future. One should however be aware of the complexity of the work and how difficult it is to produce good forecasts of the mentioned species some years ahead. The Multispecies Research Program has initiated a field of research, but continuing research in this field will be of outmost importance.

REFERENCES

Anon. 1995. Report of the Atlanto-Scandian Herring and Capelin Working Group. Copenhagen 17 - 21. Oct. 1994. ICES C.M. 1995/Assess: 9.

Bogstad, B. , I. Røttingen, P. Sandberg & S. I. Steinshamn 1994. Beskatningsstrategi for norsk vårgytende sild (in Norwegian). Foundation for Reasearch in Economics and Business Administration, Bergen. Report 84/94. 42 pp.

Flåten, O. & K. Stollery 1994. The economic effects of biological predation. Research paper 67. University of Portsmouth. 27 pp.

Nakken, O. , P. Sandberg & S. I. Steinshamn 1994. Management strategies for the Northeast Arctic cod stock. Fundation for research on Economics and Business administration. Working paper no 96/1994. 22 pp.

The Joint Fishery Commission. Protocols 1980 - 1994

Norwegian White Papers (in Norwegian):

Nr 93 (1983-83) Om retningslinjer for fiskeripolitikken

Nr 46 (1988-89) Om miljø og utvikling

Nr 32 (1989-90) Framtid i nord

Nr 32 (1990-91) På rett kjøl

Some Perspectives

By Tor Rødseth.
Department of Economics, University of Bergen, 5020 Bergen, Norway.

It is clear that the multi-species management program was a rather ambitious project. Much theoretical work is needed in both oceanography, biology, economics and statistics to form a solid basis for such large scale modelling. Furthermore, enormous estimation problems are encountered in different parts of the models, calling for masses of data and a variety of statistical methods. Thus, the enterprise may seem like trying to solve an infinite task in a 5-years research program.

However, from an other angle, one may say that this type of modelling is a way to sum up and combine present knowledge in the different areas of fishery research. Thus, the program may be seen as a milestone on a never-ending path. Central, as most of the topics investigated in the program are, it goes without saying that research on most of the issues treated will continue in the future.

This calls for a pragmatic approach: instead of brooding over the fact that we do not have a "perfect model" (which amounts to have a replica of the real world), one may pose questions like:

- What are the main weaknesses in the program?
- Where is knowledge most lacking?
- Where is uncertainty most crucial?
- What are the data needs?
- What are the costs and benefits of obtaining important data?
- What mistakes are made in a single-species approach?
- What are the costs and benefits of aggregating into more compact models?
- etc.

Such questions may be asked about different parts of the program.

OCEANOGRAPHY ETC.

From the point of view of explaining and predicting the development of fish stocks, several factors of environmental nature are important. They may work directly or indirectly, or both, in relation to fish stocks. For a given stock, temperature may work indirectly by influencing the behaviour of predating stocks, competing stocks and prey stocks, and also directly on the behaviour of the stock in question. Similar considerations can be made about other meteorological and environmental factors, like wind, currents, salinity etc.

In the Barents sea, plankton has usually been considered the main "exogenous" food supply. The term exogenous is used to indicate that feedback from other fish stock seems to be taken little into account. From the article by Giske and Skjoldal and Slagstad it seems that plankton may be divided into two groups: those which are drifting purposelessly by the currents into the area, and those which migrations may be explained by their adaptation to environmental factors.

From a management point of view, the main task is to produce research results which are of management use in this area. It is quite clear that to cover all these direct and indirect effects in a numerical model is a formidable task, in particular if one wants to use the model for prediction purposes. Research in different fields within this area is going on in different research groups which have been related to the multi-species program. And it is to be expected that this will be the case for many years to come. Parts of this research, like oceanographic and meteorological modelling, are very computer intensive and costly in other ways. It seems that such research is only efficient within a system of international co-operation.

In other areas, research groups in Norway can play a significant independent role, based on the country's long and important traditions of research. Surveying and investigating stocks of plankton and fish has been going on for many decades organised by the Institute of Marine Research. It seems that it is possible to simulate inflow and distribution of plankton with reasonable accuracy, granted that meteorological and other factor inputs are satisfactory covered. Thus, in this area of research a long run perspective is needed, since basic input information can be given only within a short time horizon. For example, weather forecasts today are limited to 4-5 days. However, much research is put into forecasting of general climatic trends. One can expect that this research will be of great value in fishery related research.

FISH MODELS

To evaluate the perspectives in the area of constructing multi-species fish models, one may take as a starting point the "balance account" for a given fish stock, in a multi-species setting:

Increase in biomass	=	Recruitment
	+	Growth
	–	Predation
	–	Fishing mortality
	–	"Natural mortality".

The last term may be taken as a proxy for factors not included in the model. That means that if residuals in a model behaves in a random way, this indicates that other factors can be left out. If, on the contrary, residuals vary systematically, this indicates that important variables or other elements are left out.

The general problem, then, is what the possibilities are to estimate and predict these factors. Except for predation, this, of course, has been the core of research in fishery biology for decades. There is a large bulk of research, both theoretical and empirical, in this area. Still, important white spots are left to be explored.

Traditionally, recruitment has been explained by the size of the spawning stock of the given species, following the pioneering work by Beverton and Holt, and of Ricker. In empirical work , recruitment functions often are not very precisely estimated. In more recent research the role of temperature for recruitment and survival of larvae has been investigated. In many cases exogenous temperature input in fish models is based on ad hoc assumptions, e.g. sinus curves. Moreover, present research on the possibility to predict general tendencies in temperature may increase the possibility to predict recruitment. At this stage it is only a matter of prediction for a very few years ahead.

Somewhat similar considerations may be made about growth. It appears that growth functions are influenced by temperature and other environmental factors. This is taken into account e.g. in MULTSPEC. Still, considerable empirical work is needed in order to establish these relations on a firm basis.

Predation, of course, is the pivotal element in multi-species models. Considerable effort has been put into investing stomach content, in particular for cod. For other species, like mammals, information about stomach content is scarce, for obvious reasons. Information about stomach content for cod exists for a number of years. Thus, one can follow the development and the changes over time in the diet.

Ideally, one should have a testable theory for explaining the diet habits of different species, perhaps along the lines of economic consumer theory - based on the utility and costs of obtaining different food elements. Such a theory does not seem to exist at present. In the absence of this, models usually are based on simple ad hoc assumptions about predation. For instance, the diet is divided on different food element in a way proportional to the supply, taking into account the spatial overlapping between predator and prey. For mammals, studies based on energetics have been made in order to assess the necessary quantum of food.

It goes without saying that to cover all the possible interactions (cannibalism included) with stomach data sufficient to give precise estimates of diet habits, is an overwhelming task. One may perhaps ask whether there are simpler and less costly ways to cover predation in the models. And in cases like mammals one may have to come up with other solutions than to collect stomach data. It seems that theoretical studies may open up new avenues in the research on predation.

Fishing mortality in principle is a decision variable for management. An efficient use of the models rests on correct reporting of catches. In practice reporting of catches is often imprecise, to say the least. Besides under-reporting, by-catches and discarding are often not reported or misreported. Furthermore, in many cases mortality is measured as the relation between catch and stock size, so that correct information on stock sizes is also needed.

In a multi-species setting, all kinds of mistakes, misreporting etc. on one species effects the information on other species. That means, for instance, that mistakes made on the assessment of stocks of commercially uninteresting species may have implications for the management of more important species, through different forms of species interactions.

On natural mortality, little more need to be said in addition to what was said above. In principle, all predation will be considered as natural mortality, in a single species setting. It may be said that most of research in fishery biology aim at substituting natural mortality with explained causes, the final aim being that all variations except random fluctuations are explained.

ECONOMICS

Like in biology, multi-species models in bio-economics of fisheries has a short history. Much of the work has been on the theoretical side, establishing conditions for long-run dynamic equilibria, and investigating optimality conditions. And much of the work has been done by economists in isolation, without input from and feedback to biological research.

The Multi-species program represents one of the first major attempts at combining biological and economic research in this area. The compact biological model AGGMULT is integrated with the economic model ECONMULT. The latter is a generalised fleet model which is compatible with many biological models, provided that definitions etc. are harmonised.

One may think of the biological part giving a starting point both for biological developments and economic considerations. Then, economic decisions give a feedback to the biological model. The two models work together in a sequential fashion.

The economic model is limited to the bioeconomic aspects of harvesting. That is, other activities in the fishing industry like processing and marketing are not included. Within this limitation two lines of development in the research may be foreseen. The first is work to improve the quality of the model in itself, i.e. obtaining better data and better estimates of parameters in the model. The other avenue is to make other uses of the model than have hitherto been made.

To discuss the first issue, one has to look at the ingredients of the harvesting model. These are elements one finds in most short or long run economic models. Factor input (effort), which may be measured in different ways, determine catches through a production function of Cobb-Douglas type with stock size as an argument. In addition to estimation problems within the Cobb-Douglas framework, other types of production functions may be relevant to explore. Moreover, production functions should in principle be estimated for each combination of vessel group (partly determined by gear) and each stock. Thus, further coverage of the production structure implies much work both on econometrics and data handling.

Catches multiplied by prices constitute gross revenue in harvesting. In a harvesting model it is natural to use first-hand prices, although they to a great extent are influenced by market prices. The two options at hand by introducing market information are (1) fixed prices and (2) demand curves with decreasing prices corresponding to increasing catches. The former approach implies correct and relevant price information, while the latter implies considerable econometric work estimating demand functions.

Thus, what has so far been done on production structure and on market information can be improved in several ways. In addition, information on cost structures for the different fleet groups are needed. Together with revenue, profits and contribution margins are calculated, and cost data may also be improved. One issue which may be pursued is the problem of evaluation of labour costs. The alternative cost of labour may be considered lower in many fisheries districts than in other parts of the country. Thus, experiments on this assumption may be of interest.

While a number of scenarios and calculation has been made within this model framework, the possibilities are far from exhausted. In some scenarios total quotas are given. From this, one may calculate the most cost-effective distribution of effort between fleet groups, which may be compared to actual distributions.

In some scenarios made, some rudimentary elements of behaviour in the fleet groups are introduced, like assuming no effort in fleet groups with negative profit margins. It is clear that much can be done to include behavioural relations in this type of models. This can, for instance, be coupled to the dynamics of the model. It is often believed that in the fishing industry, one can endure many bad years, because, when good years arrive, they much more than compensate for the former. Such behaviour can be explored in the models.

Some relatively minor extensions of the ECONMULT model can be made, which are of great interest from management point of view. Since the different fleet groups can be located , both by the participant's economic background and regionally, many distributional analyses of interest can be made within the framework of the model. This may be done in different ways. One possibility is to try to analyse the distributional consequences of a cost-effective, or social optimal management of the fleet, given the total quotas. Another is to try to estimate the social costs of different distributional goals.

As mentioned above, the complete picture of the fishing industry includes both harvesting, processing and marketing. From the point of view of the national economy, the social income of the fisheries can be thought of as "last-hand value" minus the combined costs of harvesting, processing and marketing. This gives an argument for trying to include processing and marketing in models like ECONMULT. It goes without saying that this would be a rather large project. On this point, the difference between the Norwegian fishing industry and other countries' industry is of interest. In some countries, the industry is very much integrated, which means that all the three activities are included in single firms. In Norway, the industry is little integrated. In many instances, lack of co-ordination between the three activities is demonstrated, for instance by proposing goals for Norwegian export of fish products which seem to be inconsistent with resource conditions. Thus, granted that the lack of integration will continue, it certainty is of interest to study both efficiency and distributional aspects of the organisation of the industry even if this is a large research project in itself.

MANAGEMENT

What is here meant by management, is research which lies between what has been reported on ecology, fish models and economics on the one side, and management decision problems on the other. Generally speaking , output from biological and economic research may not be in a form which is suitable for management decision making processes. In the first place, preferences by managers seem to be multi-dimensional. That is, decisions are usually not made on the basis of one single variable, like for instance, stock size or present value of resource rent. Furthermore, the dynamic aspects of different decisions are not necessarily covered by simple discounting mechanisms. The whole history of a given target variable may be of interest. And, finally, uncertainty may be handled in many different ways, not always based on standard statistical measures. For stock sizes or catches, for instance, one may be interested in such consequences of decisions as bad years or runs of bad years.

The SCENARIO BARENTS SEA project, as reviewed in the article by Scweder et. al. is an attempt at establishing a framework for such a general analysis of policy as indicated above. In principle, the scenario model comprises relevant biology and economics, the main task being to analyse the consequences of management regimes and policy decisions. This is done by a number of different performance measures. These performance measures cover a variety of possible indicators which managers may be thought to be interested in.

The SCENARIO BARENTS SEA project is carried further in the new program Management of marine resources, sponsored by the Norwegian research council. At the close of the multi-species program, the possibilities in the model framework were far from explored. Scenarios were limited to the biology, and just some few performance measures were calculated. It is to be expected that further work will include economic performance measures, and also that more types of measures will be tested out.

The preliminary reactions from management on this project seem to indicate that it has grasped essential elements of the typical management decision situation. A special feature of the project is that extensive use is made of informants from research groups and from management, qualified guesswork etc. Thus, the project may serve as a foundation for a meeting ground where insights from biologic and economic research and management experience can be pooled together in a fruitful way.

FINAL REMARK

This book is concerned with modelling. As indicated above, modelling has served as a fruitful meeting ground between biologists, economists, mathematicians and statisticians. It should also be added that the Multi-species program has also included research in political science and sociology. While this research has not been in a form which has fitted into a model frame, it goes without saying that this research has also added to our insight in problems of multi-species management in the fishing industry.